Student Study Guide

FUNCTIONS MODELING CHANGE

A PREPARATION FOR CALCULUS

STUDENT STUDY GUIDE

Raymond J. Cannon
Baylor University

Josephine J. Cannon
Baylor University

to accompany

Functions Modeling Change

A Preparation for Calculus

Third Edition

by

Eric Connally
Harvard University Extension

Deborah Hughes-Hallett
University of Arizona

Andrew M. Gleason
Harvard University

et al.

John Wiley & Sons, Inc.

COVER PHOTO © Scott Berner / Picture Quest

To order books or for customer service please, call 1-800-CALL WILEY (225-5945).

This material is based upon work supported by the National Science Foundation under Grant No. DUE-9352905. Opinions expressed are those of the authors and not necessarily those of the Foundation.

ISBN-13 978-0-470-10559-7

Printed in the United States of America

10 9 8 7 6 5 4 3 2 1

Printed and bound by Bind-Rite Robbinsville

TO THE STUDENT

How to use this Study Guide

The purpose of this Study Guide is to help you use your textbook effectively, and to be successful in your study of precalculus. We believe that to do well in a mathematics course you need to attend class and be actively involved in class. However, probably just as important is the time you spend studying and working problems outside of class. In this Study Guide are some suggestions for using your study time efficiently. This guide can promote effective study habits by:

(1) *Encouraging you to be an active, involved reader of your textbook.* Each section begins with the feature **READING YOUR TEXTBOOK**. In this feature we guide you through the reading of the corresponding section in the textbook, calling your attention to key ideas, terms, definitions, notation, and formulas. We repeatedly remind you to work through the examples with pencil and paper and to use your calculator or computer. When particular algebraic skills are required, we refer you to the appropriate sections where they are developed. Reading your textbook in this way will require time and effort, but the understanding you acquire will enable you to complete the assigned homework problems more efficiently.

(2) *Asking you to reflect on and apply what you have read.* In the **REVIEWING THE BASICS** feature, we begin with "**You should be able to**...". As you read this list, you should feel reasonably confident that you understand what you are being asked to do and that you have understood the relevant explanations and examples in your textbook. To give you immediate feedback, we include a few **Practice Problems** for you to try on your own in the space provided in the guide. These problems are usually straightforward applications of definitions and formulas. The **Solutions to Practice Problems** appear at the end of the problem set for each section.

(3) *Giving you a system for completing the exercises assigned by your instructor.* This feature is labeled **MASTERING CONCEPTS AND SKILLS** because it is by completing all the assigned homework problems that you achieve the mastery and confidence needed to do well on exams. We suggest the **√**, **?**, * system for homework problems. Ours is similar to the system used in the Student Study Guide prepared by Beverly Michael and Peg McPartland to accompany Calculus: Single Variable, by Hughes-Hallett, Gleason, et al. Space is provided for you to list the exercises assigned by your instructor. As you work on each problem, list the problem number in the space which best describes how you did on that problem. The symbols represent:

√ (check) I understand and have completed this problem.

? (question) I made progress on this problem, but either I'm not sure that what I've done is correct, or I'm not sure how to complete it.

* (starred) I really did not know how to do this problem. I need to get help on this problem or ask about it in class.

If you have studied your textbook and class notes carefully, many of the assigned problems should go into the √ category, and most should fall into either the √ or **?**

categories. If you find that you are consistently putting several problems in the * category, then you may need to spend more time studying the textbook and working through the examples before attempting the exercises, or you may need to put more effort into working on the exercises. It is important for you to get something on paper and do some thinking about every assigned problem. The more you have thought about a problem, the better you will understand the solution when it is presented in class. As you complete a problem that you placed in either the **?** or * category, circle the problem number and then rewrite it in the √ category. Your goal should be to get *every* assigned problem into the √ category within one or two more class periods. This method will allow you to see which problems gave you difficulty and thus you can study more efficiently for exams.

We hope this guide enriches precalculus for you, and that your success in this course encourages you to continue your mathematical studies.

Jo Cannon
Ray Cannon

CONTENTS

1 FUNCTIONS, LINES, AND CHANGE 1

2 FUNCTIONS 19

3 EXPONENTIAL FUNCTIONS 36

4 LOGARITHMIC FUNCTIONS 47

5 TRANSFORMATIONS OF FUNCTIONS AND THEIR GRAPHS 59

6 TRIGONOMETRIC FUNCTIONS 77

7 TRIGONOMETRY 99

8 COMPOSITIONS, INVERSES, AND COMBINATIONS OF FUNCTIONS 116

9 POLYNOMIAL AND RATIONAL FUNCTIONS 124

10 VECTORS AND MATRICES 142

11 SEQUENCES AND SERIES 152

12 PARAMETRIC EQUATIONS AND CONIC SECTIONS 161

CHAPTER ONE

FUNCTIONS, LINES, AND CHANGE

1.1 FUNCTIONS AND FUNCTION NOTATION

READING YOUR TEXTBOOK: Read Section 1.1, pp. 2-6.

As you read:

- Learn the definition of **function**. (See box, page 2.)
- Recognize four ways to present a function: **Words, Table, Graph, Formula.** (See Example 1, p. 2.)
- Understand **function notation Q =** $f(t)$ introduced on page 4, and used in Examples 2, 3, and 4 on pages 4 and 5. Notice especially that the parenthesis does not mean multiplication as it does in algebra. It will help if you practice saying aloud "f of t" when you see the notation $f(t)$. Thus in Example 4, you can say "f of 1 equals 1.8"
- Understand when a relationship is not a function, both when the relationship is given by data (See Example 5, page 5), or graphically (See Example 6, page 6).

REVIEWING THE BASICS

You should be able to:

- Decide if a relationship given by words is a function.
- Decide if a relationship given by a table is a function.
- Decide if a relationship given by a graph is a function
- Decide if a relationship given by a formula is a function.
- Use function notation to describe the relation between an input value and an output value.

Practice Problems

1. Let F be the temperature (in degrees Fahrenheit) and t the time of day. Explain why F is a function of t, but t is not usually a function of F.

---

2. Use the table to decide if P is a function of M. Explain your reasoning.

M	5	10	15	20	25
P	103	97	88	103	91

Use the same table to decide if M is a function of P. Explain your reasoning.

---

3. Explain carefully why this graph does not give W as a function of Q.

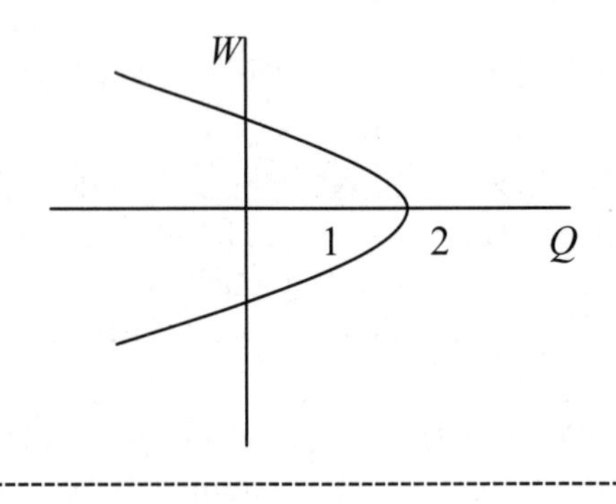

---

4. Using the table in Practice Problem 2, we can write $P = f(M)$ to mean "P is a function of M." So the notation $f(15) = 88$ means when $M = 15$ then $P = 88$. Use the table to fill in the blanks.

$f(10) =$ ________ $\qquad$ $f(\_\_) = 91$

5. The graph of a function given below shows $f(2) = 10$. Use the graph to fill in the blanks.

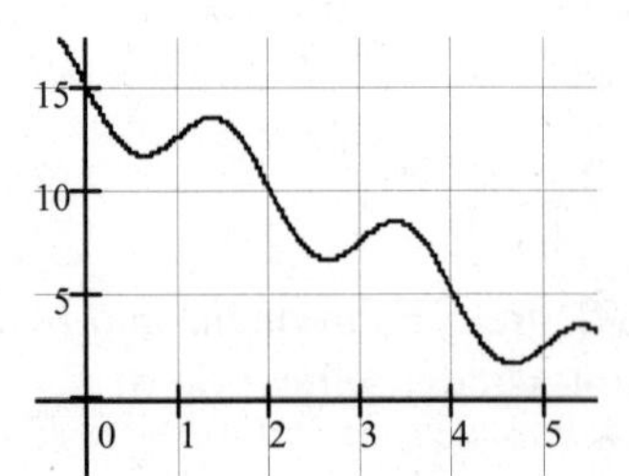

$f(0) =$ ______ $f(4) =$ _____

Solutions to Practice Problems

1. F is a function of t because at any time t there is exactly one temperature F. But t is not a function of F because knowing the temperature does not tell you the time. The same temperature F could occur at more than one time t.
2. Yes, P is a function of M because for each value of M there is exactly one value of P. No, M is not a function of P because when $P = 103$ then the value of M is not uniquely determined; M could be 5 or 20.
3. This graph fails "the vertical line test". The line $Q = 1$ intersects the graph at more than one point. Note that to give a complete answer we should specify a particular vertical line.
4. $f(10) = 97$ $f(25) = 91$
5. $f(0) = 15$ $f(4) = 5$

MASTERING CONCEPTS AND SKILLS

Use the √ , ?, * system on the exercises assigned by your instructor for this section.

Assigned Problems:

√

?

---

*

1.2 RATE OF CHANGE

READING YOUR TEXTBOOK: Read section 1.2, pp. 10-14.

As you read:

- Learn the definition of **average rate of change** (box, p. 11).
- Learn the definition of **increasing function** and **decreasing function** (box, p. 11).
- Understand the **Δ notation** for the change in a quantity. (Introduced on p.11)

- Understand average rate of change as the slope of a line (See the dashed line segments in Figures 1.12 and 1.13).

REVIEWING THE BASICS

You should be able to:

- Compute the average rate of change of a function on an interval.
- Decide by looking at the graph of a function if it is increasing on an interval, decreasing on an interval, or neither.
- Make the connection between an increasing function and the sign of the average rate of change on every interval.
- Make the connection between a decreasing function and the sign of the average rate of change on every interval.
- Use the Δ notation for change in a quantity.

Practice Problems

1. Use the graph of f to answer the questions that follow.

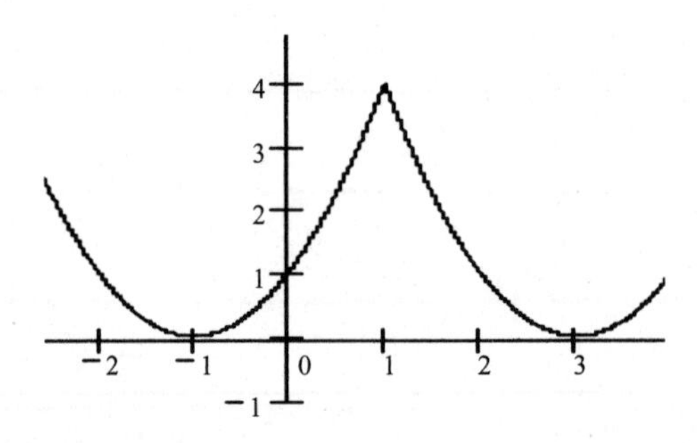

(a) $f(0) =$ ______ [Remember (x, y) on the graph means $y = f(x)$.]

(b) $f(3) =$ ______

(c) Find the average rate of change of f over the interval from $x = -1$ to $x = 1$.

(d) Find the average rate of change of f over the interval $-1 \le x \le 3$.

(e) Is f increasing on the interval $-1 \leq x \leq 0$? Explain how you arrived at your answer.

(f) Is f decreasing on the interval from $x = 2$ to $x = 3$? Explain.

(g) Is f increasing on the interval from $x = -2$ to $x = 3$? Explain.

(h) The average rate of change of f on the interval $-1 \leq x \leq 2$ is__________, which is a positive number. Is f increasing on this interval? Explain your answer.

2. If $f(x) = x^2 + 3$, then

(a) what is Δf as x changes from $x = 1$ to $x = 5$?

(b) what is $\dfrac{\Delta f}{\Delta x}$ as x changes from $x = 1$ to $x = 5$?

Solutions to Practice Problems

1. (a) $f(0) = 1$ since the point (0,1) is on the graph.
   (b) $f(3) = 0$ since the point (3,0) is on the graph.
   (c) $\dfrac{f(1) - f(-1)}{1 - (-1)} = \dfrac{4 - 0}{2} = \dfrac{4}{2} = 2$
   (d) $\dfrac{f(3) - f(-1)}{3 - (-1)} = \dfrac{0 - 0}{4} = 0$
   (e) Yes, the graph is rising from left to right on this interval.
   (f) Yes, the graph is falling from left to right on this interval.
   (g) No, because the graph is falling on the interval from $x = -2$ to $x = -1$, and from $x = 1$ to $x = 3$. So f is neither increasing nor decreasing on the interval from $x = -2$ to $x = 3$.
   (h) $\dfrac{f(2) - f(-1)}{2 - (-1)} = \dfrac{1 - 0}{3} = \dfrac{1}{3}$, which is positive, but f is not increasing on the interval from $x = -1$ to $x = 2$ because the graph is falling from $x = 1$ to $x = 2$.

2. (a) $\Delta f = f(5) - f(1) = 5^2 + 3 - (1^2 + 3) = 24$;

(b) $\frac{\Delta f}{\Delta x} = \frac{f(5) - f(1)}{5 - 1} = \frac{24}{4} = 6$.

MASTERING CONCEPTS AND SKILLS

Use the √ , ?, * system on the exercises assigned by your instructor for this section.

Assigned Problems:

√

?

---

*

1.3 LINEAR FUNCTIONS

READING YOUR TEXTBOOK: Read section 1.3, pp. 17-24.

As you read:

- Notice that a **constant rate of change** in Examples 1and 2 gives rise to a linear function.
- Notice the **sign of the rate of change**. The rate of change is **positive** in Examples 1 because the population is **increasing**. The rate of change is **negative** in Example 2 because the value is **decreasing**.
- Notice the **units of the rate of change**. For example, the population growth rate is Example 1 is measured in **people per year**; that is, units of P per unit of t. Observe that in the other examples, the rate of change is measured in **units of output per unit of input.**
- Understand the verbal pattern for a linear function given above the box on p. 20.
- Know the significance of the parameters b (the y-intercept) and m (the slope) in the linear equation $y = mx + b$. (See box, p. 20 and Example 3.)

- Examples 4 and 5 show two ways to recognize when a table of values could represent a linear function: if the inputs increase in equal steps (increments), check for equal changes in outputs (see Example 4). Otherwise, check for constant rate of change (see Example 5).

- Use your graphing calculator to follow along in Example 6. The shape of a graph can depend on the viewing window, as in Figures 1.21 and 1.22.

REVIEWING THE BASICS

You should be able to:

- Decide if a table of values could represent a linear function.
- Write the formula for a linear function if you know the initial value and the rate of change.
- Find the rate of change of a linear function given two data points.
- Interpret the slope of a line as a rate of change.
- Recognize an equation of the form $y = mx + b$ as linear with initial value b and rate of change m.

Practice Problems

1. Fill in the blanks: A linear function has a _____________ rate of change. The graph of a linear function is a _____________. If $f(x) = mx + b$, then f is a _____________ function. The graph of f is a ________. The parameter m gives the ____________ of the graph; b is the _______________.

2. In each table below, x changes by equal increments, $\Delta x =$ ____. Fill in the blanks with the changes in the y-values.

x	$f(x)$	Δf
5	2	
		___
10	6	
		___
15	10	
		___
20	14	

x	$g(x)$	Δg
5	2	
		___
10	6	
		___
15	8	
		___
20	9	

For which function do the y-values increase by equal increments? _____

Which function could be linear? ____ What is the rate of change of the linear function? __________ Plot the points given in each table. What do you see that supports your conclusion?____

3. A car that cost $19,200 in 1998 is depreciating at a rate of $2500 per year. Write a formula for the value V of the car t years from 1998. (Use $t = 0$ in 1998. Remember that "depreciating" means decreasing in value.)

   $V =$ ___________________________

4. The following table shows the cost for an organization to purchase different numbers of specially designed T-shirts.

Number of shirts	20	50	100	200
Cost (dollars)	165	360	685	1335

   (a) Show that the cost appears to be a linear function of the number of shirts.

   (b) Explain in practical terms what the slope represents in this situation.

5. What is the y-intercept of the graph of $y = 3x - 500$? ________ Explain why the standard viewing window [-10,10]x[-10,10] does not show any points on this graph.

Solutions to Practice Problems

1. A linear function has a **constant** rate of change; the graph of a linear function is a **line**. If $f(x) = mx + b$, then f is a **linear** function. The graph of f is a **line**. The parameter m gives the **slope** of the graph; b is **the y-intercept.**

2. $\Delta x = 5$; $\Delta f = 4$, constant, so f is linear. The rate of change is $\frac{4}{5}$. The points for f lie on a line; the points for g do not. The successive changes in g are 4, 2, and 1.

3. $V = 19200 - 2500\,t$.

4. (a) $\frac{360-165}{50-20} = \frac{685-360}{100-50} = \frac{1335-685}{200-100} = 6.5$

   (b) The points lie on a line with slope 6.5. The rate of change is \$6.5 per shirt; that is,
   each additional shirt costs \$6.50.

5. The y-intercept is -500, which is not in the standard viewing window. The largest y-value is -470 when $x = 10$, so none of the points are in the standard window.

MASTERING CONCEPTS AND SKILLS

Use the √, ?, * system on the exercises assigned by your instructor for this section.

Assigned Problems:

√

?

---

*

1.4 FORMULAS FOR LINEAR FUNCTIONS

READING YOUR TEXTBOOK: Read Section 1.4, pp. 27-31.

As you read:

- Notice how you use a data point to determine the vertical-intercept b once you have found the slope m. In Example 1 the data is given in a table. In Example 2, the data is given in a graph. In both cases you find the rate of change, m, as you did in Section 1.3.

- Study Example 3 carefully. Notice in part (a) that the equation is first written in words before any letters are used as variables. Writing "verbal models" like this one will help you think clearly about what each term in your equation represents.

- Example 3, part (a), also shows how to change an equation of the form $Ax + By = C$ into the form $y = mx + b$.

- Notice the interpretation of the intercepts and the slope of the graph in part (b) of Example 3.

- Learn the **point-slope form** of a linear equation. Example 4 shows how using this form lets you write down an equation for the line without first finding the y-intercept.

- Recognize the three different ways to write an equation of a line. (See box p.31)

- Be sure you follow all the computations in each example. It is a good idea to work them out yourself with pencil and paper and your calculator.

REVIEWING THE BASICS

You should be able to:

- Find a formula for a linear function using two data points. The points may be given in a table, in a graph, or in function notation.

- Use the **point-slope form** to write a linear equation if you know the slope and a point on the line.

- Rewrite a linear equation in the **slope-intercept form**, and identify the slope and the vertical intercept.

- Find a formula for a linear function from a verbal description.

Practice Problems

1. Identify the slope and the y-intercept of the graph of each equation.
   (a) $y = 3x + 7$ (b) $5x - 2y = 8$

2. Find a formula for the linear function with the given properties.

   (a) Slope -2 and y-intercept 10 (b) Passes through $(2, -5)$ and $(-1, 4)$

3. Use the point-slope form to write a formula for the linear function having slope $\frac{3}{4}$ and passing through the point (6, –2).

4. A sorority paid \$165 for 20 T-shirts and \$360 for 50 T-shirts. Assuming the cost is a linear function of the number of shirts, find a formula for the cost function.

5. Find a formula for the linear function that has the following graph.

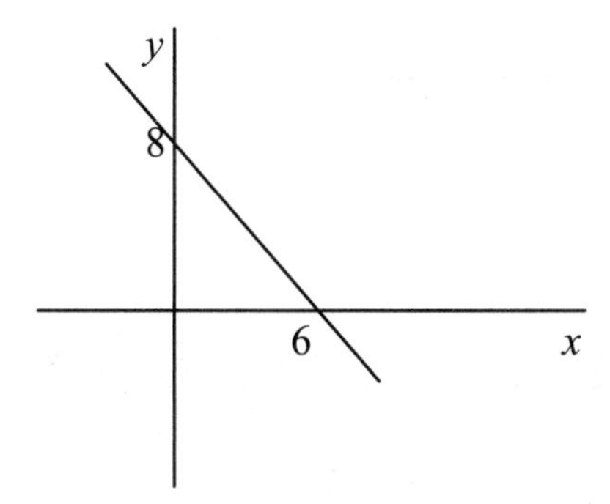

Answers to Practice Problems

1. (a) slope 3; y-intercept 7 ; (b) Solving for y, we get $-2y = 8 - 5x$; $y = -4 + 2.5x$; so slope is 2.5; y-intercept is –4
2. (a) $y = -2x + 10$; (b) $m = \dfrac{-5-4}{2-(-1)} = \dfrac{-9}{3} = -3$; Using the slope-intercept form and

the point (–1,4), we get $4 = -3(-1) + b$; so $b = 1$; and $y = -3x + 1$. You should get the same answer using the point (2, –5), or using the point-slope form with either point.

3. The equation is $y - (-2) = \frac{3}{4}(x - 6)$, or $y = -2 + \frac{3}{4}(x - 6)$
4. Since the function is linear, $C = mx + b$, where C = cost and x = number of shirts. The two points are (20,165) and (50, 360), so $m = \frac{360-165}{50-20} = 6.5$. Using (20,165) in the point-slope form, $C - 165 = 6.5(x - 20)$, or $C = 6.5x + 35$. (Check to see that you get the same answer using the point (50,360)
5. The points are (0,8) and (6,0), so $m = \frac{8-0}{0-6} = \frac{-4}{3}$. We are given the y-intercept, so the equation is $y = \frac{-4}{3}x + 8$.

MASTERING CONCEPTS AND SKILLS

Use the √ , ?, * system on the exercises assigned by your instructor for this section.

Assigned Problems:

√

?

*

1.5 GEOMETRIC PROPERTIES OF LINEAR FUNCTIONS

READING YOUR TEXTBOOK: Read section 1.5, pp. 35-40.

As you read:

- Learn what the parameters m and b tell you about the line. (See box, p. 36) Examples 1 and 2, pp. 35-36, illustrate these effects.

- Be sure you follow all the computations in **finding the point of intersection of two lines** in Example 3. Do the arithmetic to show that the coordinates of the point satisfy both equations.

- Graph the three functions in Example 3 on your calculator. Use a viewing window so that your graphs looks like Figure 1.33. Be sure you follow the computations for finding the points of intersection, and trace to find these points on your graph.

- Learn the **equations of horizontal lines and vertical lines**. Understand the difference between a slope of zero (horizontal line) and no slope (vertical lines). (Box, p. 39)

- Learn the relationships between slopes of **parallel lines**, and between slopes of **perpendicular lines**. (Box, p.39)

REVIEWING THE BASICS

You should be able to:

- Determine by looking at the equation $y = mx + b$ if the line is rising or falling, and if the y-intercept is positive or negative.
- Compare the slopes of two lines that are graphed on the same set of axes.
- Determine from their equations whether two lines are parallel, perpendicular, or neither.
- Write an equation of a line parallel or perpendicular to a given line and which passes through a given point.
- Find the point of intersection of two lines by solving a system of equations.
- Write equations of vertical and horizontal lines.

Practice Problems

1. Each line has an equation of the form $y = mx + b$. For each line, tell whether m is positive, negative, or zero, and whether b is positive, negative, or zero.

(a)

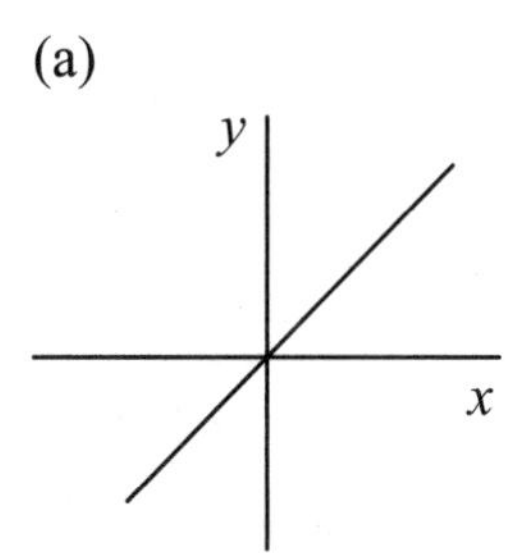

(b)

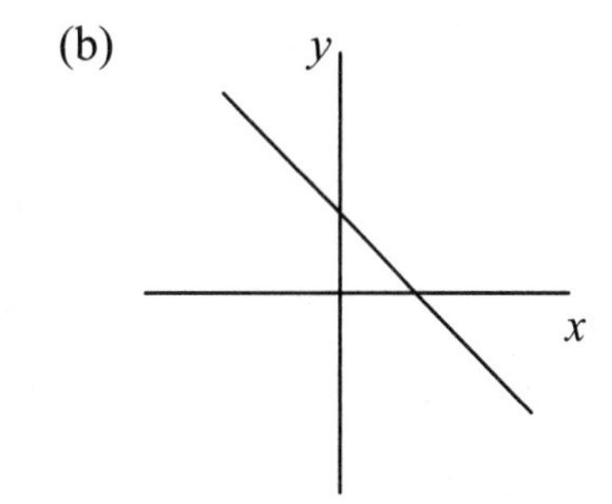

(c)

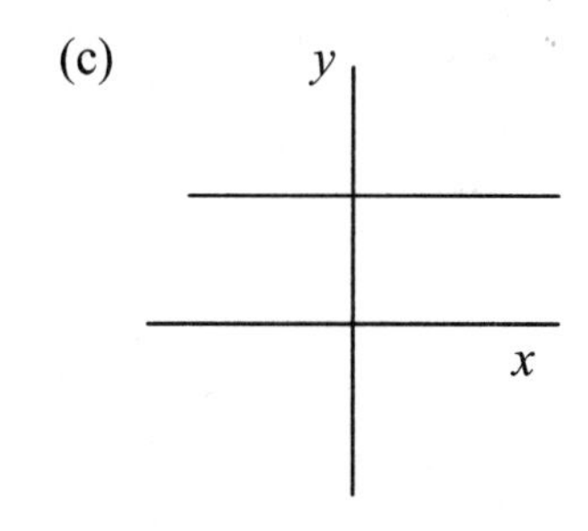

2. Arrange the slopes of the lines in ascending order. Use m_1 for the slope of line l_1, and so on.

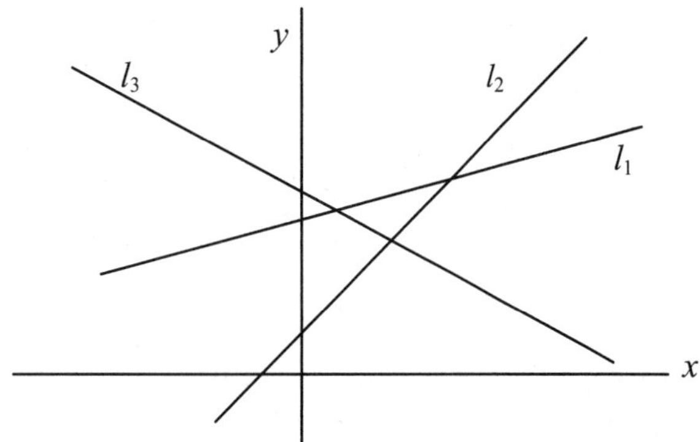

3. Rewrite each equation in the slope-intercept form, and determine if the line is parallel or perpendicular to the line $y = 2x - 5$.

   (a) $6x - 3y = 8$

   (b) $2x + 5y = 3$

   (c) $4x + 8y = 5$

4. Write an equation of the line through the point (2, –3) which is

   (a) parallel to the line $y = \frac{4}{5}x + 1$

   (b) perpendicular to the line $y = \frac{4}{5}x + 1$

5. Use algebra to find the point of intersection of the lines $2x - y = 8$ and $6x + 2y = 3$. One way to do this is to write each equation in the slope-intercept form, and proceed as in Example 3, p. 37. Check your solution by graphing each line on your calculator and finding the point of intersection.

6. Write an equation of the vertical and horizontal lines through the point (2, -3).

   Vertical ________________ Horizontal ______________

Solutions to Practice Problems

1. (a) $m > 0, b = 0$; (b) $m < 0, b > 0$; (c) $m = 0, b > 0$.

2. Since $m_3 < 0$ and the others are positive, m_3 is the smallest. Since l_2 is steeper than l_1, the order is $m_3 < m_1 < m_2$

3. (a) $y = 2x - \frac{8}{3}$; parallel: (b) $y = \frac{-2}{5}x + \frac{3}{5}$; neither; (c) $y = \frac{-1}{2}x + \frac{5}{8}$; perpendicular
4. (a) $y - (-3) = \frac{4}{5}(x-2)$, or $y = -3 + \frac{4}{5}(x-2)$; (b) $y = -3 - \frac{5}{4}(x-2)$
5. In slope-intercept form, $y = 2x - 8$ and $y = -3x + 1.5$. For the y-values to be equal, we need $2x - 8 = -3x + 1.5$, or $5x = 9.5$, so $x = 1.9$, and $y = 2(1.9) - 8 = -4.2$. The point of intersection is (1.9, –4.2).
6. Vertical: $x = 2$; Horizontal: $y = -3$

MASTERING CONCEPTS AND SKILLS

Use the √ , ?, * system on the exercises assigned by your instructor for this section.

Assigned Problems:

√

?

---

*

1.6 FITTING LINEAR FUNCTIONS TO DATA

READING YOUR TEXTBOOK: Read Section 1.6, pp. 44-47

As you read:

- Become familiar with the terms **scatter plot** and **linear regression**.
- Use your calculator or computer to compute the regression line for the viscosity using the data from Table 1.34, p.44.
- Understand the terms **interpolation** and **extrapolation**, and understand which tends to be more reliable.
- Understand what a **correlation coefficient** measures. (See Figure 1.58 and Example 2)

REVIEWING THE BASICS

You should be able to:

- Make a scatter plot of a set of data and draw a regression line by eye.
- Estimate the correlation coefficient by eye.
- Use a calculator or computer to find the equation of a regression line.
- Interpret the slope and each intercept of the line.
- Interpolate and extrapolate using the equation, and discuss the reliability of the results.

Practice Problems

1. Match the r-values with the scatter plots. $r = 1$ $r = -.7$ $r = 0$

(a)

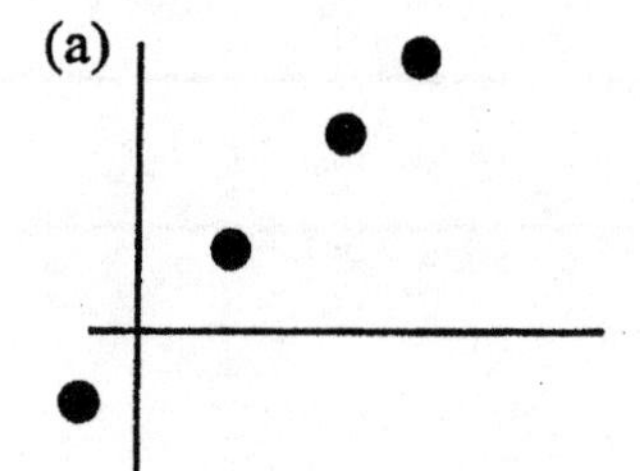

(b)

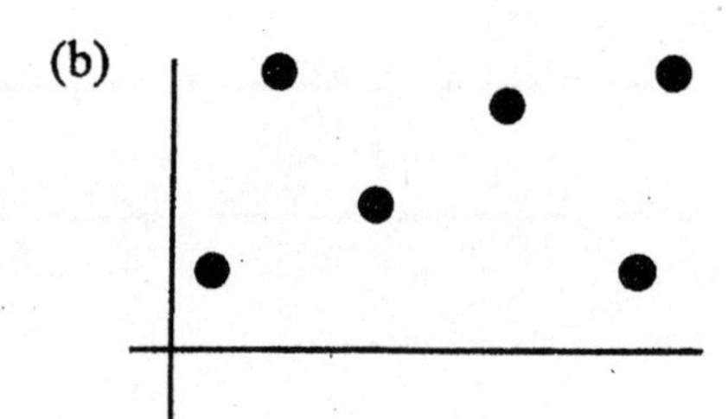

(c)

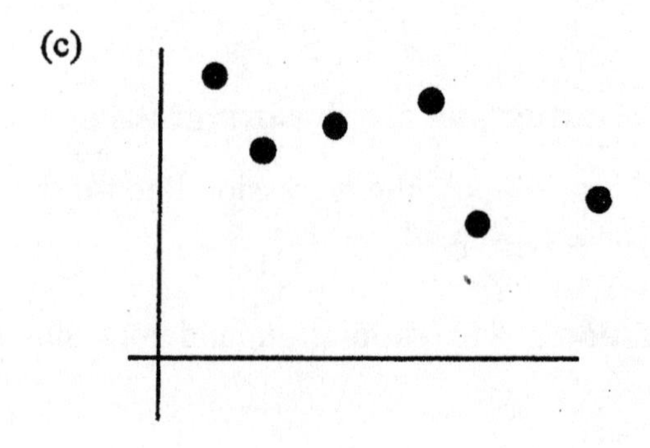

2. Use the table below to answer the questions.

x	7.5	8.4	10.3	12.8
y	15.2	17.6	20.4	24.3

(a) Make a scatter plot of the data.
(b) Draw an estimated regression line by eye.
(c) Use a calculator or computer to find the equation of the regression line.
(d) Use the equation to predict the value of y when $x = 14.0$

Solutions to Practice Problems

1. (a) $r = 1$; (b) $r = 0$; (c) $r = -.7$

2. (a) and (b)

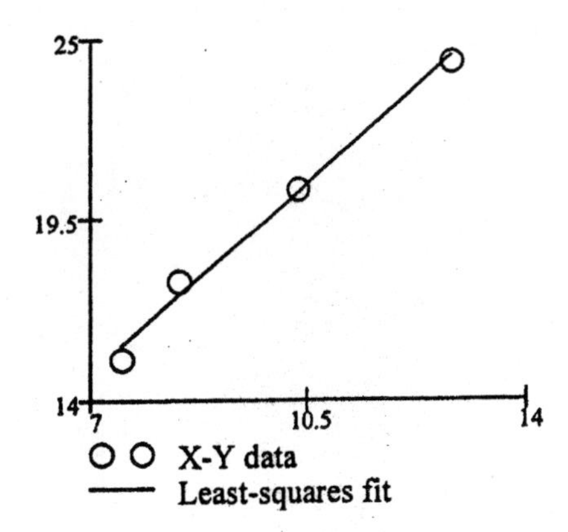

(c) $y = 3.2 + 1.7\,x$; (d) 27.0

MASTERING CONCEPTS AND SKILLS

Use the √ , ?, * system on the exercises assigned by your instructor for this section.

Assigned Problems:

√

?

---

*

CHAPTER TWO

FUNCTIONS

2.1 INPUT AND OUTPUT

READING YOUR TEXTBOOK: Read Section 2.1, pp. 62-65.

As you read:

- Study the use of function notation carefully in Examples 1– 4, pp. 62-63. In Example 4 notice the difference between $h(a-2)$ in part (b) and $h(a)-2$ in part (c). Compare parts (b) and (d) to see the difference between $h(a-2)$ and $h(a)-h(2)$. Remember that the parentheses do not mean multiplication as they do in algebra.

- Work carefully through the algebra in Example 4. In part (b), remember that $(a-2)^2$ means $(a-2)(a-2)$, so you get the "middle term" $-4a$.

- Be aware of the difference between **simplifying an expression** and **solving an equation.** In Examples 4, you are using algebra to simplify the expression you get when evaluating a function. In Examples 5-7, you are using algebra to solve an equation.

- Work carefully through the algebra in Examples 5-7.

- In Examples 8 and 9, notice the difference between **evaluating** $f(b)$, where b is a given input, and **solving** the equation $f(x) = b$, where b is a given output. These questions do not require algebra, but rather an understanding of function notation, and how a table or a graph can define a function.

REVIEWING THE BASICS

You should be able to:

- Evaluate a function for a given input, using a formula, table, or graph.
- Use algebra to simplify expressions obtained when evaluating a function using a formula.

- Solve an equation of the form $f(x) = b$ for the input, using a formula, table or a graph. For some functions given by a formula, you need to use algebra to solve the equation. (For more complicated formulas, you may need a graphing calculator or computer to solve the equation.)

- Interpret the statement $f(a) = b$ when the function models a relationship between physical quantities.

Practice Problems

1. Let $f(x) = 3x - 5$.

   (a) Find $f(0)$.

   (b) Solve $f(x) = 0$.

   (c) Evaluate and simplify $f(1/a)$.

2. Use the graph of f given below to estimate:

   (a) $f(2)$ ____ (b) The value(s) of x where $f(x) = 2$ __________

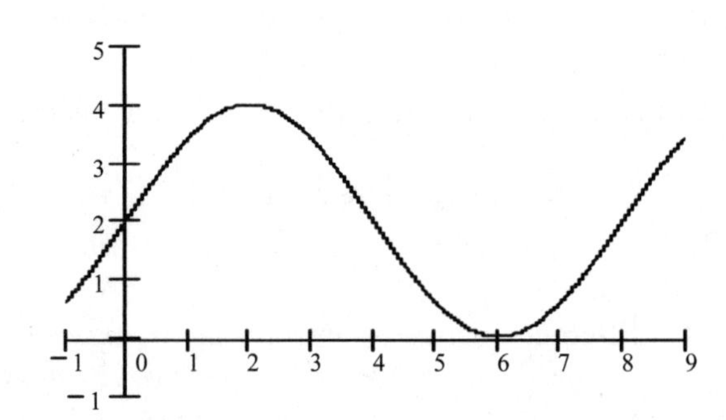

3. Let $f(x) = x^2 + 3$. Evaluate and simplify

   (a) $f(a-2)$

   (b) $f(a) - f(2)$

   (c) $f(a) - 2$

4. Let $T = f(R)$ be the temperature predicted by the chirp rate of the snowy tree cricket; see Table 1.1 on page 3 of the text.

   (a) Evaluate and interpret $f(80)$.

   (b) Solve and interpret $f(R) = 80$.

Solutions to Practice Problems

1. (a) -5; (b) $3x - 5 = 0$ when $3x = 5$, so $x = 5/3$; (c) $3(1/a) - 5 = 3/a - 5a/a = (3-5a)/a$
2. (a) $f(2) = 4$; (b) $f(x) = 2$ when $x = 0$ or 4 or 8
3. (a) $f(a-2) = (a-2)^2 + 3 = a^2 - 4a + 4 + 3 = a^2 - 4a + 7$
   (b) $f(a) - f(2) = a^2 + 3 - (2^2 + 3) = a^2 - 4$; (c) $f(a) - 2 = a^2 + 3 - 2 = a^2 + 1$
4. (a) $f(80) = 60$, which means that when the cricket is chirping at 80 chirps per minute, the predicted temperature is 60 degrees.
   (b) Solving $f(R) = 80$ for R gives $R = 160$. This means that the chirp rate that predicts a temperature of 80 degrees is 160 chirps per minute.

MASTERING CONCEPTS AND SKILLS

Use the √ , ?, * system on the exercises assigned by your instructor for this section.

Assigned Problems:

√

?

*

2.2 DOMAIN AND RANGE

READING YOUR TEXTBOOK: Read Section 2.2, pp. 69-71.

As you read:

- Learn the definitions of **domain** and **range**. (See box, p.69.)
- Understand why the domain of a function that models a physical situation may not be the same as the domain of the abstract algebraic function. This is illustrated in Examples 1 and 2.
- Use your calculator or computer to sketch the graph of the function in Example 3. Use a viewing window so your graph looks like the one in Figure 2.8 on p. 70. Trace along the graph, and notice that the y-values get near, but are less than, 260. This example should help you understand that the range of a function is the set of outputs, or the y-values on the graph.

- Note in Examples 4 and 5 illustrate two things you will need to avoid in determining the domain of a function given by an algebraic formula: getting zero as a denominator and getting a negative number under a square root sign.

- Use your calculator or computer to graph the functions in Examples 3, 4, and 5, and understand how the graphs show agreement with the conclusions arrived at by examining the formulas algebraically. But also understand how unthinking reliance on the graph can lead to incorrect conclusions. For example, tracing the function in Example 5 cannot tell you that the function is defined for *all* $x > 4$. The calculator may show a smallest x-value, a little larger than 4. You need to think about the formula to determine the exact domain.

- It will help to remember, when trying to determine range, that just as zero has no reciprocal, zero is not the reciprocal of any number. So $1/x$ can never equal 0 and hence 0 is not in the range of g in Example 4. In general, determining range is a harder problem than finding domain. Example 5 helps you think about the possible outputs to determine range. When you use the formula to determine the range, it is a good idea to graph the function and see if the graph agrees with your conclusions. Example 3 shows how the graph (in an appropriate window) can suggest the range.

REVIEWING THE BASICS

You should be able to:

- Determine the domain and range of a function by looking at the graph.
- Determine the domain and range of a function by examining the formula.
- Determine the domain and range of a function by the constraints of the situation being modeled.

Practice Problems

1. The complete graph of a function is shown below. Determine the domain and range of the function.

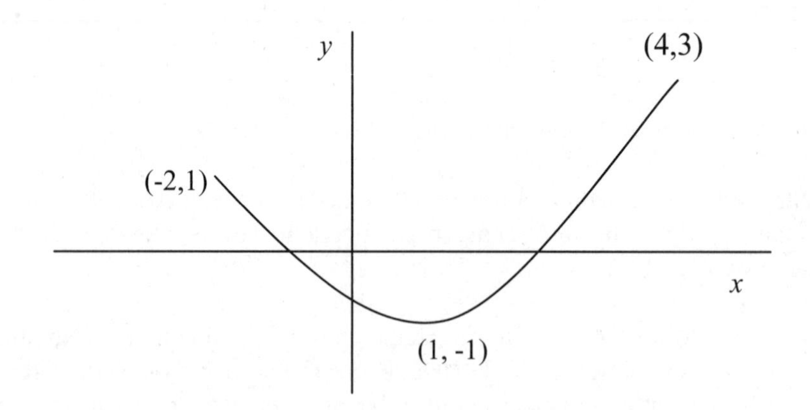

Domain ________________ Range ______________

2. Let $f(x) = \sqrt{5-x}$, $g(x) = \dfrac{1}{x+3}$, and $h(x) = x^2 + 1$

(a) Determine the domain of each function.

(b) Use algebra to show that 2 is in the range of each function. (Find the value of x so that the output of the function is 2.)

(c) Is 0 in the range of f? ______ of g?_____ of h?_____. If so, give the value of x that produces 0 as the output.

(d) Use your calculator or computer to sketch a graph of each function, and determine the range of each function.

3. Explain the difference between the domain of the abstract function $f(x) = x^2$ and the domain of the function $A = f(x) = x^2$ which gives the area of a square as a function of the length of the side.

Answers to Practice Problems

1. Domain $-2 \le x \le 4$; Range $-1 \le y \le 3$
2. (a) $5 - x \ge 0$ when $5 \ge x$, so the domain of f is the interval $(-\infty, 5]$. The domain of g is all real numbers except -3. The domain of h is all real numbers.

   (b) $\sqrt{5-x} = 2$ when $5 - x = 4$, so when $x = 1$. So 2 is in the range of f.

   $\dfrac{1}{x+3} = 2$ when $x + 3 = ½$, or when $x = -2.5$. So 2 is in the range of g.

   $x^2 + 1 = 2$ when $x^2 = 1$, or when $x = 1$ or -1. So 2 is in the range of h.

   (c) $\sqrt{5-x} = 0$ when $x = 5$, so 0 is in the range of f. Zero is not the reciprocal of any number, so 0 is not in the range of g. Since $1 + x^2 \ge 1$ for every real number x, it follows than 0 is not in the range of h.

   a. The range of f is the interval $[0, \infty)$.
   The range of g is all real numbers except 0.
   The range of h is the interval $[1, \infty)$
3. The domain of the abstract function $f(x) = x^2$ is all real numbers, but if x represents the length of the side of a square, then x must be positive, so the domain is $(0, \infty)$.

MASTERING CONCEPTS AND SKILLS

Use the √, ?, * system on the exercises assigned by your instructor for this section.

Assigned Problems:

√

?

---

*

2.3 PIECEWISE DEFINED FUNCTIONS

READING YOUR TEXTBOOK: Read Section 2.3, pp. 73-76.

As you read:

- Learn the **definition of the absolute value of x**, written $|x|$. (See box, p. 76.) Be able to write it *exactly* as it appears in the box. You may have trouble with the $-x$, because you already know that the absolute value of a number is not negative. But putting a minus sign in front of a number changes the sign. So when $x < 0$, $-x > 0$.

- Learn what is meant by a **piecewise defined function**. The graph in Figure 2.10, p. 73, is the graph of one function. The graph is in two different looking "pieces" because the formula is different in different parts of the domain.

- Notice the use of the open circle to show that the point is not on the graph, and the large dot to show that the point is on the graph. (See Figure 2.11, p.74.)

- Study Examples 2 and 3, p. 74-75 for physical situations modeled by a piecewise defined function.

REVIEWING THE BASICS

You should be able to:

- Write the definition of $|x|$ for any real number x.
- Rewrite a function involving absolute values without using absolute value bars.
- Sketch the graph of a piecewise defined function.
- Write a piecewise function that models a physical situation.

Practice Problems

1. Fill in the blanks. Give exact values, not decimal approximations.

$|-5| =$ _____ $\qquad |\sqrt{5}-2| =$ _________ $\qquad |\pi - 4| =$ _______

2. Let

$$f(x) = \begin{cases} x-5 & \text{for } x \le 3 \\ 1 & \text{for } 3 < x < 5 \\ x+1 & \text{for } x \ge 5 \end{cases}$$

(a) Fill in the blanks.

$f(-1) =$ _____ $\qquad f(3) =$ ____ $\qquad f(3.1) =$ ____

$f(4.9) =$ ____ $\qquad f(5) =$ ____ $\qquad f(27) =$ ____

(b) Sketch the graph of f for $-1 \leq x \leq 7$. Use open circles and large dots to make clear what happens at $x = 3$ and $x = 5$.

3. Sara has a job in sales that pays a base salary of $200 per week plus a commission of 20% of her weekly sales in excess of $1000.

   (a) How much did Sara make in a week if her sales were:

   $950? _______ $1000? ________ $2500? _______

   (b) Write a function which gives Sara's salary S in terms of her weekly sales x.

   (c) Sketch the graph.

Answers to Practice Problems

1. $|-5| = -(-5) = 5$; $|\sqrt{5} - 2| = \sqrt{5} - 2$ since $\sqrt{5} - 2 > 0$; $|\pi - 4| = -(\pi - 4) = 4 - \pi$ since $\pi - 4 < 0$.
2. (a) $f(-1) = -1 - 5 = -6$; $f(3) = 3 - 5 = -2$; $f(3.1) = 1$; $f(4.9) = 1$; $f(5) = 5 + 1 = 6$; $f(27) = 27 + 1 = 28$

(b)

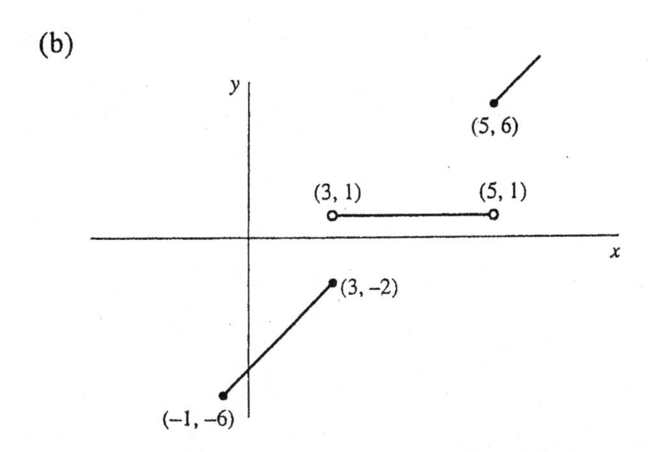

3. (a) Sales of \$950, salary is \$200; sales of \$1000, salary is \$200; sales of \$2500, salary is \$200 plus 20% of \$1500, so her salary is $200 + .2(1500)$, or \$500.

   (b) $S(x) = \begin{cases} 200 & \text{for } 0 \le x \le 1000 \\ 200 + .2(x - 1000) & \text{for } x > 1000 \end{cases}$

   (c)

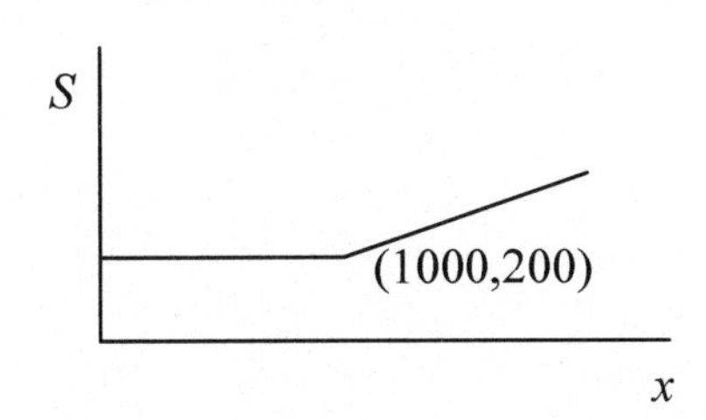

MASTERING CONCEPTS AND SKILLS

Use the √ , ?, * system on the exercises assigned by your instructor for this section.

Assigned Problems:

√

?

*

2.4 COMPOSITE AND INVERSE FUNCTIONS

READING YOUR TEXTBOOK: Read Section 2.4, pp. 79-82

As you read:

- Learn what **composition of functions** is, and how you can follow one function evaluation with another one.
- Pay careful attention to Example 3, which shows the order of the composition matters.
- Learn what an **inverse function** does. This idea is explained in the paragraph just before Example 4, and is illustrated in Example 4.
- Pay special attention to the **warning** about the notation f^{-1}, p. 80. Example 4 (b) shows that $f^{-1}(x)$ *does not mean* the reciprocal: $\dfrac{1}{f(x)}$.
- Notice the different **units** attached to the input 4 in Example 4. In part (a), the input 4 represents the number of years since 2007, while in part (b) the input 4 represents thousands of birds.
- Be sure to follow the algebra in Example 5, where given a formula for a function f you find a formula for the inverse function f^{-1}.
- Understand the relationship between the functions f and f^{-1}, as shown in Example 6

REVIEWING THE BASICS

You should be able to:

- Be able to identify when composition of functions is being used.
- Given formulas for $f(x)$ and $g(x)$, be able to compute $g(f(x))$.
- Interpret the statement $f^{-1}(b) = a$ when the function f models physical quantities. Your answer should include units.
- Evaluate f^{-1} from a table of values or a graph of f itself.

Practice Problems

1. Let $f(x) = 5 - 2x$, and $g(x) = 1/x^2$. Evaluate the following and simplify your answers.

   (a) $f(3x)$ (b) $3f(x)$ (c) $f(x+3)$

   (d) $f(x) + 3$ (e) $f(g(x))$

2. Let $C(x)$ be the cost, in thousands of dollars, of building a house with area x square feet. Match each story below to one of the expressions that folow.
   (a) I figured out the cost of my house, and then added $10,000 for landscaping. ___
   (b) After figuring out the cost of my house, I got a raise and decided to increase the size of my house by 1000 square feet. ____
   (c) After figuring out the cost, I decided to reduce cost by 25%. ____

   (i) $C(x+1000)$ (ii) $C(x+10{,}000)$ (iii) $C(x)+10$ (iv) $C(x) - .25$ (v) $.75C(x)$

3. Let $P = f(t)$ be the population, in millions of people, of a country t years from 2005. Interpret each statement.

   (a) $f(10) = 8$ ____

   (b) $f^{-1}(12) = 18$ ____

4. Use the following graph of f to fill in the missing values.

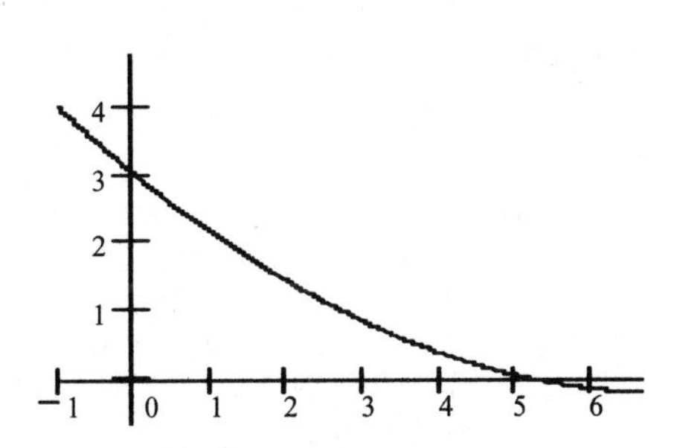

   (a) $f(0) =$ ______ (b) $f^{-1}(0) =$ _______

Solutions to Practice Problems

1. (a) $f(3x) = 5 - 2(3x) = 5 - 6x$; (b) $3f(x) = 3(5 - 2x) = 15 - 6x$; (c) $f(x+3) = 5 - 2(x+3) = -1 - 2x$; (d) $f(x) = 5 - 2x + 3 = 8 - 2x$; (e) $f(g(x)) = 5 - 2g(x) = 5 - 2(1/x^2) = (5x^2 - 2)/x^2$
2. (a) (iii); (b) (i); (c) (v)
3. (a) In 2015 the population will be 8 million people. (b) The population will reach 12 million in 18 years from 2005, or in 2023.
4. (a) $f(0) = 3$; (b) $f^{-1}(0) = 5$

MASTERING CONCEPTS AND SKILLS

Use the √ , ?, * system on the exercises assigned by your instructor for this section.

Assigned Problems:

√

?

---

*

2.5 CONCAVITY

READING YOUR TEXTBOOK: Read Section 2.5, pp. 84-86.

As you read:

- Be sure to follow the examples, and see how the definitions in the box on p.86 apply to each figure.
- Notice in Table 2.17 that the rates of change become *less negative* so the curve's descent is slowing down; the graph goes down but bends up.
- Study Example 2 to see how a decreasing rate of change produces a graph that is concave down.
- Be sure to understand that concavity deals with the way a graph is bending; the graph can be either increasing or decreasing. Be familiar with the four cases presented in the graphs given in Figures 2.22-2.25.

REVIEWING THE BASICS

You should be able to:

- Recognize when a graph is increasing or decreasing, and concave up or concave down on an interval.
- Make the connection between increasing rate of change and concave up, and between decreasing rate of change and concave down.

Practice Problems

1. Label each graph as either increasing or decreasing, and as concave up or concave down.

(a)

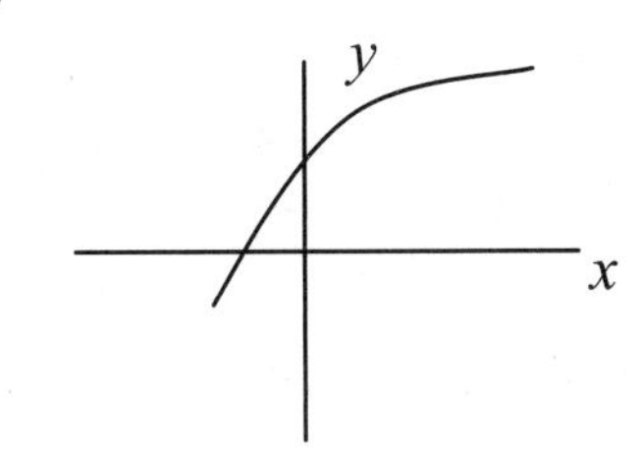

(b)

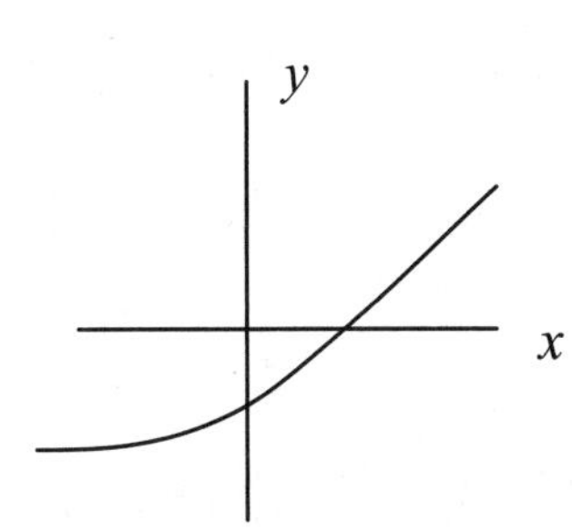

(c)

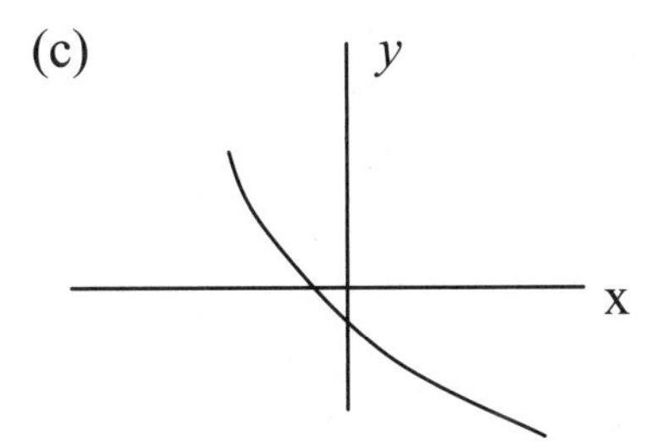

(d)

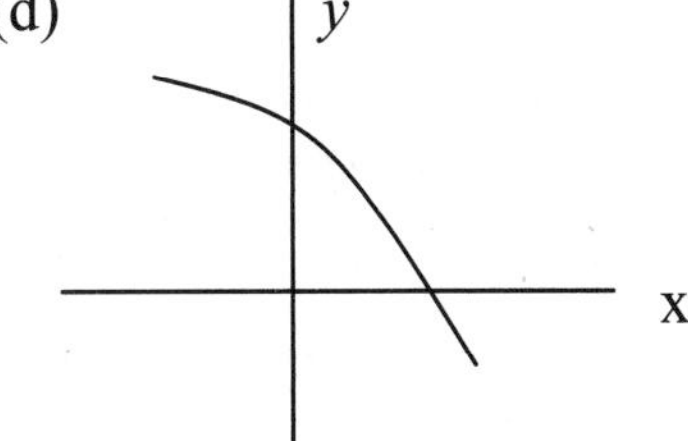

2. Each of the functions given in the table is increasing, but each increases in a different way. Match the functions to the graphs. Explain your reasoning.

x	$f(x)$	$g(x)$	$h(x)$
0	4.0	40	20
10	5.2	48	30
20	6.4	60	38
30	7.6	72	44
40	8.8	86	48

(a)

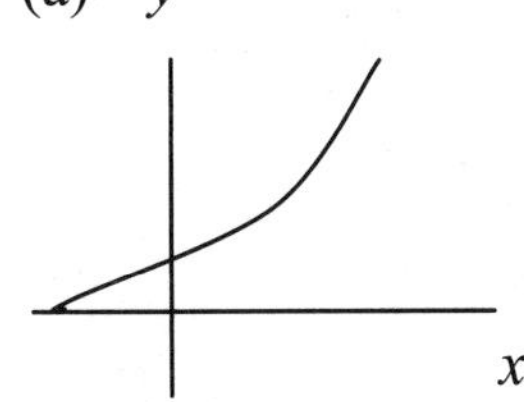

(b)

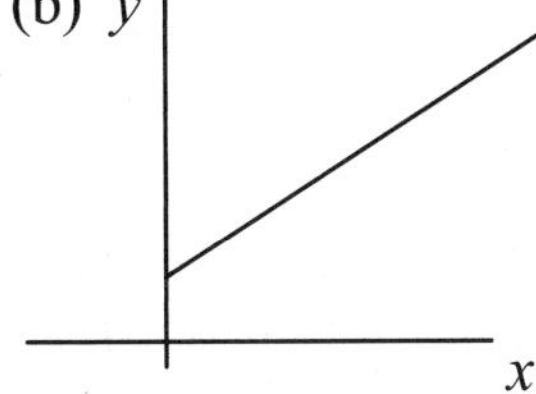

(c)

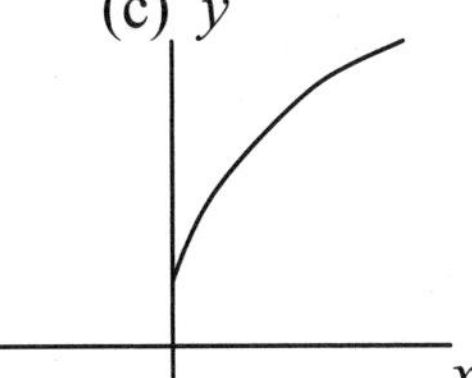

Answers to Practice Problems

1. (a) increasing, concave down; (b) increasing, concave up; (c) decreasing, concave up; (d) decreasing, concave down
2. $\Delta x = 10$ is constant for all functions. For f, $\Delta f = 1.2$, constant, so f is linear, graph (b). Successive increases in g are 8, 10, 12, and 14. Since the increases are getting larger, the graph of g is concave up, graph (a). The increases in h are 10, 8, 6, and 4. Since the increases are getting smaller, the graph is concave down, graph (c)

MASTERING CONCEPTS AND SKILLS

Use the √ , ?, * system on the exercises assigned by your instructor for this section.

Assigned Problems:

√

?

---

*

2.6 QUADRATIC FUNCTIONS

READING YOUR TEXTBOOK: Read Section 2.6, pp. 88-91.

As you read:

- Learn the general form of a **quadratic function**, and form a mental picture of a **parabola**.
- Remember the distinction between evaluating $f(0)$ and solving $f(x) = 0$.
- **Finding the x-intercepts** (zeros) requires solving the equation $f(x) = 0$, either by **factoring** (Example 1) or using the **quadratic formula** (Example 2). You should memorize the quadratic formula if you have forgotten it; it is found in Example 2 and again on pages 93 and 102.

- If you need more examples in solving quadratic equations, see Examples 8-11 on pages 101 and 102. Note especially Examples 9 and 10 which each address a common mistake.

- Read the graph in Figure 2.28 carefully; you are not looking at the path of the ball. The y-coordinate of a point on the graph gives the height of the ball at a particular time. Read the graph in Example 5 in the same careful way.

- Note that the concavity of the parabola is determined by the coefficient of x^2. If this number is positive, the parabola is concave up; if negative, the parabola is concave down.

REVIEWING THE BASICS

You should be able to:

- Find a formula for a parabola that has two given x-intercepts.

- Find the zeros (x-intercepts) of a quadratic function by factoring or using the quadratic formula.

- Recognize the distinction between $-\sqrt{9}$, which is the same as –3, and $\sqrt{-9}$, which is not a real number.

- Describe the type of concavity of a parabola.

Practice Problems

1. Let $f(x) = x^2 - 4x - 5$.

   (a) Find the x-and y-intercepts.

   (b) Sketch the graph. Is the graph concave up or concave down?

2. Find a possible formula for the parabola. Is the graph concave up or concave down?

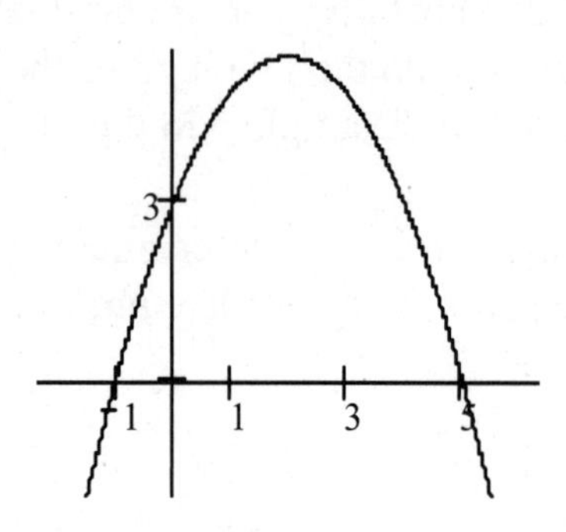

3. Try to use the quadratic formula to find the zeros of $f(x) = x^2 - 4x + 5$. What do you conclude about the graph of f? Is the graph concave up or concave down?

Solutions to Practice Problems

1. (a) $f(0) = -5$ is the y-intercept. Factor to find the x-intercepts: $x^2 - 4x - 5 = (x - 5)(x + 1) = 0$ when $x = 5$ or $x = -1$.

   (b)

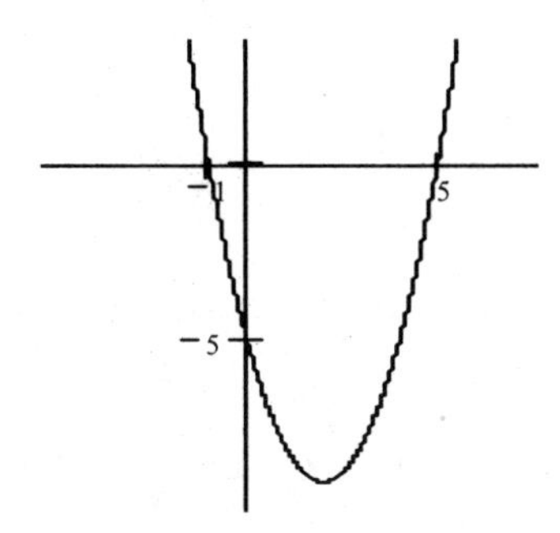

2. (a) $y = a(x + 1)(x - 5)$; $y = 3$ when $x = 0$, so $3 = a(1)(-5) = -5a$; $a = -.6$. The equation is $y = -.6(x + 1)(x - 5)$

3. Using the quadratic formula, we have $x = (4 \pm \sqrt{16 - 4(5)})/2 = (4 \pm \sqrt{-4})/2$, which is not a real number. The graph of f has no x-intercepts. Since the coefficient of x^2 is 1, which is positive, the parabola is concave up.

MASTERING CONCEPTS AND SKILLS

Use the √ , ?, * system on the exercises assigned by your instructor for this section.

Assigned Problems:

√

?

---

*

CHAPTER THREE

EXPONENTIAL FUNCTIONS

3.1 INTRODUCTION TO THE FAMILY OF EXPONENTIAL FUNCTIONS

READING YOUR TEXTBOOK: Read Section 3.1, pp. 106-112.

As you read:

- Notice in Examples 1 and 2 that when a quantity grows by a fixed *percentage* each year, the increases themselves increase, resulting in graphs that curve upward. Compare this type of growth with linear growth, where the quantity grows by a fixed *amount* each year. In Example 3, the quantity decreases by a fixed percentage per year, so the decreases get smaller, producing a graph that is falling but bending upward.

- Pay special attention to the section on Growth Factors and Percent Growth Rates, pp. 108-109. Understand that increasing the old salary by 6% is the same as multiplying it by the factor $1 + 0.06$, or 1.06. This is the key to understanding exponential growth. In the same way, decreasing a quantity by 11.4% is the same as multiplying it by $1 - 0.114$, or .886.

- Follow the computations on pp.109-110 to see how growth by a fixed percentage per year gives rise to an exponential function.

- Understand the information summarized in the box, p. 110, and the graphs in Figures 3.4 and 3.5.

- Work through the computations in Examples 4-8, pp.111-112. Be sure you are getting the correct answers when you use your calculator to evaluate the exponential expressions.

REVIEWING THE BASICS

You should be able to:

- Write an exponential function to model a quantity that is growing (or decaying) by a fixed percentage in a given time period.

- Use your calculator to evaluate exponential expressions.
- Determine the percentage growth rate from the formula for an exponential function.

Practice Problems

1. The population of a country in 1995 was 53.4 million and growing at a rate of 2.3 % per year. Assuming that the same growth rate continues, write a formula for the population t years from 1995, and use the formula to predict the population in 2005.

2. A car purchased in 1998 for \$21,500 is depreciating at a rate of 15% per year. Write a formula for the value of the car t years from 1998, and use the formula to predict the value of the car in 2003.

3. The population (in millions) of a state is given by the formula $P(t) = 6.3(1.02)^t$, where t is the number of years from 1990.
   (a) What was the population in 1990?

   (b) What is the annual percentage growth rate?

Solutions to Practice Problems

1. $P(t) = 53.4(1.023)^t$; $P(10) = 53.4(1.023)^{10} = 67.03$ million
2. $V(t) = 21{,}500(1 - .15)^t = 21{,}500(.85)^t$; $V(5) = 21{,}500(.85)^5 = 9539.66$ dollars
3. (a) $P(0) = 6.3$ million; (b) The base $b = 1.02 = 1 + .02$, so $r = .02$, or 2% annual growth rate.

MASTERING CONCEPTS AND SKILLS

Use the √ , ?, * system on the exercises assigned by your instructor for this section.

Assigned Problems:

√

?

---

*

3.2 COMPARING EXPONENTIAL AND LINEAR FUNCTIONS

READING YOUR TEXTBOOK: Read Section 3.2, pp. 115-119

As you read:

- Notice that the test for linear or exponential functions given in the box on p. 115 applies only when x changes by equal steps. Notice also that you need to check differences and ratios for each consecutive pair of y-values.

- Work through the computations for finding a formula for an exponential function on pp.115-116. You will need your calculator. Remember to use parentheses around fractional exponents.

- Work through the computations in Example 1, p.116-117. Notice in part (a) that to find a *linear* function to fit two points, you use the *difference* between the y-values in finding the *slope*. To find an *exponential* function in part (b), you use the *ratio* of the y-values in finding the *growth factor.* In both cases you then use one of the points to find the initial value.

- Read the paragraph about the similarities and differences between linear and exponential functions. Example 2 illustrates these ideas.

- Learn the important fact about the rate of exponential growth given on p.118 and illustrated in Example 3 and Figure 3.8.

REVIEWING THE BASICS

You should be able to:

- Decide if a table of values could represent a linear function, an exponential function, or neither.

- Write a formula for an exponential function whose graph passes through two given points.

Practice Problems

1. Determine which of the functions in the following table could be exponential and which could be linear. Write formulas for the linear and exponential functions.

x	4	8	12	16	20
$f(x)$	23.6	20.8	18.0	15.2	12.4
$g(x)$	12.0	16.8	20.2	25.7	42.3
$h(x)$	12.0	16.8	23.52	32.93	46.1

2. Find a possible formula for the graph. Assume the function is exponential.

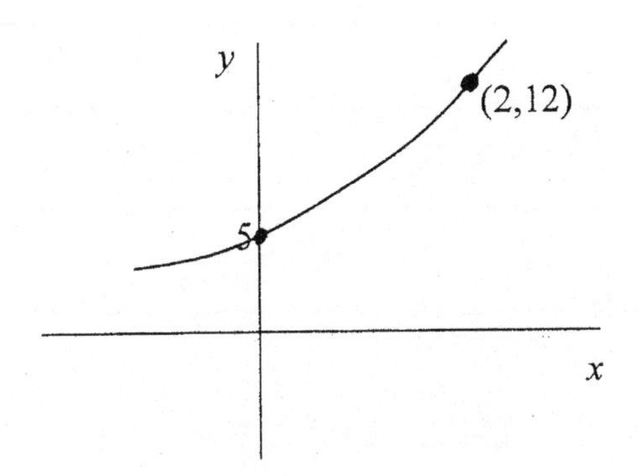

3. A radioactive substance decays at a rate of 3.5% per hour. Assume 100 grams are initially present. Write a formula for the amount Q remaining after t hours.

Solutions to Practice Problems

1. $\Delta x = 4$. Checking consecutive differences for f, we see that each $\Delta f = -2.8$. So f is linear, and the slope $m = \Delta f/\Delta x = -2.8/4 = -0.7$. Using the point (12,18), we get the equation $y - 18 = -0.7(x - 12)$, or $y = -0.7x + 26.4$. Checking differences for $g(x)$, we see that $16.8 - 12 = 4.8$, but $20.2 - 16.8 = 3.4$, so g is not linear. Checking ratios, we see that $16.8/12 = 1.4$ but $20.2/16.8 = 1.2$, so g is not exponential. Checking ratios for h, we see that $16.8/12 = 23.522/16.8 = 32.93/23.52 = 46.1/32.93 = 1.4$, so h is exponential. To find the formula for h, we know $h(x) = ab^x$. So $h(8)/h(4) = ab^8/ab^4 = b^4$, so $b^4 = 1.4$, and $b = 1.4^{1/4}$. Using the point (4,12), we get $12 = a(1.4^{1/4})^4 = a(1.4)$. So $a = 12/1.4 = 8.57$. So $h(x) = 8.57(1.4^{1/4})^x \approx 8.57(1.088)^x$.
2. $f(2)/f(0) = ab^2/a = b^2$, so $b^2 = 12/5 = 2.4$, and $b = \sqrt{2.4} \approx 1.55$. $a = f(0) = 5$. So $f(x) = 5(\sqrt{2.4})^x \approx 5(1.55)^x$.
3. $Q(t) = 100(1 - 0.035)^t = 100(0.965)^t$.

MASTERING CONCEPTS AND SKILLS

Use the √ , ?, * system on the exercises assigned by your instructor for this section.

Assigned Problems:

√

?

---

*

3.3 GRAPHS OF EXPONENTIAL FUNCTIONS

READING YOUR TEXTBOOK: Read Section 3.3, pp. 122-126.

As you read:

- Look for properties of the graphs of exponential functions. List them as you proceed through the section. How do a and b affect the graph?
- Remember that $b^0 = 1$ for any base b of an exponential function. To get the vertical intercept, we let $t = 0$, and then $Q = a\,b^0 = a$. So the parameter a gives the vertical intercept. This is illustrated in Figure 3.14, p. 123.
- We have already seen that base $b > 1$ gives an increasing function, and $0 < b < 1$ gives a decreasing function. Figure 3.15, p.123 shows you how the size of b affects the steepness of the graph.
- Use your graphing calculator or computer to graph $y = 2^x$ and $y = 3^x$ in the same window, and notice which is steeper.
- Learn the definition of **horizontal asymptote** (box, p. 124). The notation is explained in the paragraph before the box and in the box. Learn to use this notation, and say aloud what it means. The reason for having this definition at this point is to be able to discuss another important feature of the graphs of exponential functions. Figure 3.16, p.124, helps you see what it means to say that the horizontal axis is an asymptote of the graph of an exponential decay function, $0 < b < 1$.
- Note that for positive a, the graphs of exponential functions are concave up. See Figures 2.19 (exponential growth) and 2.20 (exponential decay) on pp 84-85.
- Use your graphing calculator as you work through Examples 2 and 3. Try to use the same viewing windows as seen in Figures 3.17 and 3.18. You can trace along the graph to the point that has the desired y-value and then read the corresponding x-value. You can get a more accurate answer by also graphing the functions $y = 337{,}000{,}000$ in Example 2 or $y = 25$ in Example 3 and then using the intersect feature on the calculator.

REVIEWING THE BASICS

You should be able to:

- Determine the y-intercept by examining the formula for an exponential function.
- Match formulas to graphs of exponential functions by understanding how the base affects the steepness of the graph.
- Identify the x-axis as a horizontal asymptote of an exponential function.
- Use a graphing calculator to solve an equation involving an exponential function.

Practice Problems

1. Without using a calculator, match the formula to the graph.

   (i) $y = 3(.7)^x$ _____ (ii) $y = 3(.8)^x$ ______ (iii) $y = 3(1.2)^x$ ____

   (iv) $y = 3(1.4)^x$ _____ (v) $y = 4(1.2)^x$ ______ (vi) $y = -3(.7)^x$ ____

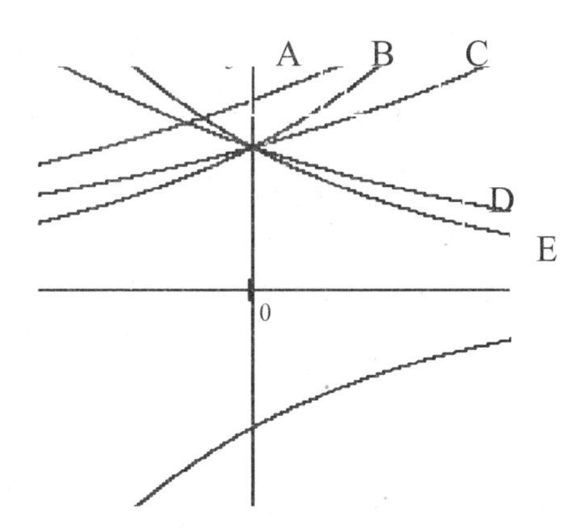

2. Solve the equation $3 \cdot 2^x = 10$ using your graphing calculator.

3. If $Q = 6 \cdot (1.03)^t$, then
   (a) Is the graph of Q increasing or decreasing?
   (b) Is the graph concave up or concave down?
   (c) Is the x-axis a horizontal asymptote of the graph? Explain.
   (d) If t is time in years, then what does 3% represent?

4. If $Q = 6 \cdot (0.97)^t$, then
   (a) Is the graph of Q increasing or decreasing?
   (b) Is the graph concave up or concave down?
   (c) Is the x-axis a horizontal asymptote of the graph? Explain.
   (d) If t is time in years, then what does 3% represent?

Answers to Practice Problems

1. (i)-(E); (ii)-(D); (iii)-(C); (iv)-(B); (v)-(A); (vi)-(F)
2. $t \approx 1.737$
3. (a) increasing; (b) concave up; (c) yes, as $x \to -\infty, Q \to 0$; (d) the annual rate of growth.
4. (a) decreasing; (b) concave up; (c) yes, as $x \to \infty, Q \to 0$; (d) the annual rate of decrease.

MASTERING CONCEPTS AND SKILLS

Use the √, ?, * system on the exercises assigned by your instructor for this section.

Assigned Problems:

√

?

---

*

3.4 CONTINUOUS GROWTH AND THE NUMBER *e*

READING YOUR TEXTBOOK: Read Section 3.4, pp. 130-133.

As you read:

- Learn that e represents a number (e = 2.71828...) that has a non-repeating decimal, just as π represents a number with a non-repeating decimal (π = 3.14159...)

- Know what is meant by **continuous growth rate.** (See box, p.131 and Examples 1-3.)

- Notice in Examples 1 and 2 that when you are given a **continuous growth rate**, your exponential model will have e as the base; Example 3 gives a **continuous decay rate.**

- Be able to distinguish between **annual** and **continuous** growth rates; see Example 4 and the discussion at the top of page 133.

REVIEWING THE BASICS

You should be able to:

- Write an exponential growth model, given a continuous growth rate.
- Rewrite $Q = a\,e^{kt}$ in the form $Q = a\,b^t$, and state the continuous percent growth rate and the percent growth rate per unit of time.

Practice Problems

1. Rewrite $N = 5\,e^{-0.47t}$ in the form $N = A\,b^t$. ________________________

2. A town of 240,000 is growing at a continuous rate of 3.5% per year.

   (a) Write a formula for the population after t years. ____________________

   (b) Find the annual percent growth rate. __________________________

3. A radioactive substance decays according to the formula $Q(t) = Q_0\,e^{-0.4t}$, where $Q(t)$ is the amount remaining after t years, and Q_0 is the original amount when $t = 0$.

   (a) What is the continuous percent decay rate? ____________

   (b) What is the annual percent decay rate? ________

Solutions to Practice Problems

1. $N = 5(e^{-0.47})^t \approx 5(.625)^t$.
2. (a) $P = 240{,}000\,e^{.035t}$ (b) $240{,}000\,e^{.035t} = 240{,}000(e^{.035})^t \approx 240{,}000\,(1.036)^t$, so the annual growth rate is 3.6%.
3. (a) Q is decaying at a continuous rate of 40 % per year
   (b) $Q = Q_0\,(e^{-0.4})^t = Q_0\,(0.67)^t$. So $0.67 = 1 + r$, and $r = -0.33$. So Q is decaying at an annual rate of 33% per year.

MASTERING CONCEPTS AND SKILLS

Use the √ , ?, * system on the exercises assigned by your instructor for this section.

Assigned Problems:

√

?

---

*

3.5 COMPOUND INTEREST

READING YOUR TEXTBOOK: Read Section 3.5, pp. 136-138.

As you read:

- Remember that increasing a quantity by a certain percentage, say 6 %, is the same as multiplying that quantity by the factor 1.06. That is, $Q + .06Q = 1.06\,Q$. (Review pp. 108-110 if you need to see again how growth by a constant percent is modeled by an exponential function.)

- Study p. 136, and be sure you understand the terms **compound interest, nominal rate,** and **effective annual yield,** or **effective rate.** See Examples 1 and 2.

- Try to *understand* the formula for compound interest (see box, p.137), rather than relying on memorizing letters. The quantity r/n is the percent growth rate per compounding period, and so $1 + r/n$ is the growth factor per period. The exponent nt is the number of compounding periods in t years. Notice that the nominal rate r is changed to decimal form in the formula.

- Do the calculation $(1 + .06/365)^{365}$. Remember to use parentheses around the base.

- Study the explanation of **continuous compounding,** pp.137-138, and learn the formula in the box, p. 138. Notice that the annual growth factor $(1 + r/n)^n \rightarrow e^r$ as $n \rightarrow \infty$; that is, as the number of compoundings per year gets larger and larger.

- Notice in Example 3 that computing $e^{.0795}$ gives you the base $b = 1 +$ effective annual rate.

- Follow the use of the formulas, and do the calculations in Example 3.

- Remember that the base $b = 1 + r$, where r is the annual percent growth rate (assuming t is measured in years). (See Example 3.)

REVIEWING THE BASICS

You should be able to:

- Answer questions using the formulas for compound interest.

- Find the effective annual yield for a nominal rate r compounded n times a year, or compounded continuously.

- Use your calculator correctly to evaluate exponential expressions. Remember to use parentheses around the exponent if the exponent is a product and around the base when the base is a sum.

Practice Problems

1. Suppose $5000 is invested at a nominal rate of 6% per year. Find the balance in the account at the end of 10 years if interest is compounded:

   (a) annually

   (b) daily

   (c) continuously.

2. How much must be invested to a nominal rate of 8% per year to have a balance of $10,000 after 5 years in interest is compounded:

   (a) monthly

   (b) continuously

3. How long will it take $2000 invested at a nominal rate of 7% per year to grow to $5000 if interest is compounded:

   (a) annually

   (b) continuously

4. Suppose an account pays a nominal rate of 9% per year. Find the effective annual yield if interest is compounded:

   (a) quarterly

   (b) continuously

Solutions to Practice Problems

1. (a) $5000(1.06)^{10} = \$8954.24$ (b) $5000(1+.06/365)^{3650} = \$9110.14$

   (c) $5000e^{.06(10)} = \$9110.59$

2. (a) $P(1+.08/12)^{60} = 10{,}000$
   $P = 10{,}000/(1+.08/12)^{60} = \$6712.10$

   (b) $Pe^{.08(5)} = 10{,}000$
   $P = 10{,}000/e^{.08(5)} = \$6703.20$

3. (a) $5000 = 2000(1.07)^t$
   $2.5 = 1.07^t$
   $t \approx 13.5$ yrs;
   With annual compounding, it will take 14 years.

   (b) $5000 = 2000\,e^{.07t}$
   $2.5 = e^{.07t}$
   $t = 13.09$ years, or about 13 years and 33 days

4. (a) In one year, $B = P(1+.09/4)^4 = P(1.093)$
   So the effective yield is 9.3%

   (b) $e^{.09} = 1.094$, so the effective yield is 9.4%

MASTERING CONCEPTS AND SKILLS

Use the √ , ?, * system on the exercises assigned by your instructor for this section.

Assigned Problems:

√

?

*

CHAPTER FOUR

LOGARITHMIC FUNCTIONS

4.1 LOGARITHMS AND THEIR PROPERTIES

READING YOUR TEXTBOOK: Read Section 4.1, pp. 152-157

As you read:

- Memorize the definition of **common logarithm**: $\log x = y$ means $10^y = x$. (See box, p. 152.)

- Practice changing a logarithm statement to the equivalent exponential statement. (See Example 1, p. 152.)

- Practice changing an exponential statement to the equivalent logarithm statement. (See Example 2, p. 152.)

- Notice the repeated use of the definition to evaluate logarithms in Example 3(a)-(e). But in part (f), you see that it does not make sense to ask "What is log(-100)?"

- Apply the "inverse properties" of $\log x$ and 10^x (see box page 153) to do the evaluations in Example 4.

- Learn the properties of logarithms (Box, p.154). Keep in mind as you learn the last three properties that a logarithm is an exponent. These properties of logarithms follow from the laws of exponents. For example, when you multiply numbers with the same base, you add the exponents. Working through the justification of these properties (pp. 156-157) is good practice using the definition of logarithm.

- Example 5 is very important. It shows you how to use logarithms to solve exponential equations. These equations occur when we are trying to answer questions about when an exponentially growing quantity reaches a certain size. (Remember you solved these equations graphically earlier). Work through all the details. Notice especially that before you take the log of both sides, you get the equation into the form $a^t = b$. Make sure you are using your calculator correctly.

- Pay special attention to the box, p. 156, and the examples following. If you are getting wrong answers, you may be making one of the common errors pointed out here.

- Learn the notation and definition of the **natural logarithm**: $\ln x = y$ means $e^y = x$. (See boxes, p. 155.) Notice that the natural logarithm uses e as the base, while the common logarithm uses 10 as the base. The properties (second box, p. 155) of $\ln x$ are the same as for $\log x$.

- Example 6 shows you how to use natural logarithms to solve equations. Rework Example 5 taking the natural logarithm of both sides. You should get the same answer.

REVIEWING THE BASICS

You should be able to:

- Write an exponential statement equivalent to a given logarithm statement.
- Write a logarithm statement equivalent to a given exponential statement.
- Use the definition of $\log x$ and $\ln x$ to evaluate certain logarithmic expressions without using a calculator.
- Use the properties of $\log x$ and $\ln x$ to rewrite expressions.
- Use properties of logarithms to solve equations involving logarithms.
- Use logarithms to solve exponential equations.
- Use your calculator to evaluate exponential and logarithmic expressions.

Practice Problems

1. Write the statement that is equivalent to the given statement.

Logarithm Statement	Exponential Statement
(a) $\log 100 = 2$	________________
(b) $\ln 8 \approx 2.079$	________________
(c) ________________	$e^2 \approx 7.389$
(d) ________________	$10^{-3} = .001$

2. Evaluate the following expressions *without* using a calculator.

(a) $\log 1$

(b) $\ln 1$

(c) $\log \dfrac{1}{100}$

(d) $\ln \sqrt{e}$

(e) $\ln e^5$

3. Use your calculator to evaluate each expression.

(a) $\ln \frac{17}{3}$

(b) $\ln 17 - \ln 3$

(c) $\frac{\ln 17}{\ln 3}$

4. Solve for t.

(a) $\log (2t + 3) = 1$

(b) $50\,(1.04)^t = 75$

(c) $28\,e^{.08\,t} = 40$

Solutions to Practice Problems

1. (a) $10^2 = 100$; (b) $e^{2.079} \approx 8$; (c) $\ln 7.389 \approx 2$; (d) $\log .001 = -3$
2. (a) $\log 1 = 0$ since $10^0 = 1$; (b) $\ln 1 = 0$ since $e^0 = 1$; (c) $\log \frac{1}{100} = \log 10^{-2} = -2$

   (d) $\ln \sqrt{e} = \ln e^{1/2} = \frac{1}{2}$; (e) $\ln e^5 = 5$
3. (a) 1.7346; (b) 1.7346; (c) 2.5789
4. (a) $\log (2t + 3) = 1$ means $2t + 3 = 10^1$, so $2t = 7$, and $t = 3.5$.

   (b) $1.04^t = 75/50 = 1.5$; Take the natural log of both sides: $\ln 1.04^t = \ln 1.5$. Use property of logarithms: $t \ln 1.04 = \ln 1.5$; $t = \ln 1.5 / \ln 1.04 \approx 10.338$.

   (c) $e^{.08\,t} = 40/28$; $\ln e^{.08\,t} = \ln (40/28)$; $.08\,t = \ln(40/28)$; $t = [\ln(40/28)]/.08$; $t \approx 4.46$

MASTERING CONCEPTS AND SKILLS

Use the √ , ?, * system on the exercises assigned by your instructor for this section.

Assigned Problems:

√

?

---

*

4.2 LOGARITHMS AND EXPONENTIAL MODELS

READING YOUR TEXTBOOK: Read Section 4.2, pp. 159-164.

As you read:

- Examples 1-7 show you how to use logarithms as a tool to solve exponential equations. These equations arise when we ask how long it takes an exponentially growing quantity to reach a certain amount. You have already solved these equations graphically (p. 125).

- Follow the computations and the justification for each step in Example 1. Notice that before you take the log of both sides, you get the equation in the form $b^t = a$. Then the property $\log b^t = t \log b$ allows you to solve for t. Do the calculations on your calculator.

- Notice in Example 3 that the variable t appears in the exponent on both sides of the equation. Follow the algebra which gets the equation into the form $b^t = a$. Be careful when you calculate the final answer. When you compute the log of a fraction, the fraction must be in parentheses.

- Learn what is meant by the terms **doubling time** (p. 161) and **half-life** (p. 162).

- Notice in Example 4 how the doubling time you found in part (a) is used to answer part (b) easily.

- Remember in Examples 6 and 7 that the half-life is the time t for which $Q(t) = \frac{1}{2} Q_0$.

  Try to justify each step in the computations, especially when properties of exponents and logarithms are used. For example, you need to use $\ln e^x = x$ in Example 7.

- Study Example 12 to see an equation with the variable in the exponent, but which cannot be solved by logarithms. This equation cannot be written in the form $b^t = a$. Find the solution by graphing, as in Figure 4.3.

- Learn how to change any exponential function to one with base e.

- Be careful of trying to memorize too much. You should have memorized the definition: $y = \ln x$ means $e^y = x$. When you know this, you won't need to memorize $k = \ln b$ when you change from base b to base e. Changing from base e to base b is just a matter of using the properties of exponents: $e^{kt} = (e^k)^t$, so $b = e^k$. (See Examples 8 and 11.)

REVIEWING THE BASICS

You should be able to:

- Use logarithms to solve exponential equations.
- Recognize when an equation cannot be solved using logarithms, and solve by graphing.
- Rewrite any exponential function $Q = a\,b^t$ in the form $Q = a\,e^{kt}$.
- Rewrite $Q = a\,e^{kt}$ in the form $Q = a\,b^t$, and state the continuous percent growth rate and the percent growth rate per unit of time.

Practice Problems

1. Solve each equation. Use logarithms to get exact solutions if possible. Check your answers by also solving graphically.

   (a) $2.6(1.04)^t = 4.1$

   (b) $2.6(1.04)^t = 4.1(1.02)^t$

   (c) $2(1.04)^t = t + 20$

2. Prices in a certain country are increasing at a rate of 12% per year. How long will it take for prices to double?

3. A radioactive substance is decaying at a rate of 12% per year. Find its half-life.

4. Find the ratio of A to B if $\log A - \log B = 2$.

5. Rewrite $Q = 6.3\,(1.28)^t$ in the form $Q = A\,e^{kt}$. ________________________

Solutions to Practice Problems

1. (a) $1.04^t = 4.1/2.6$

$\ln(1.04^t) = \ln(4.1/2.6)$

$t \ln 1.04 = \ln(4.1/2.6)$

$$t = \frac{\ln(4.1/2.6)}{\ln 1.04} \approx 11.61$$

(b) $(1.04)^t/(1.02)^t = 4.1/2.6$

$(1.04/1.02)^t = 4.1/2.6$

$\ln(1.04/1.02)^t = \ln(4.1/2.6)$

$t \ln(1.04/1.02) = \ln(4.1/2.6)$

$$t = \frac{\ln(4.1/2.6)}{\ln(1.04/1.02)} \approx 23.46$$

(c) Solve by graphing. There are two solutions. Since the exponential function must overtake the linear function, we know to look for a first quadrant solution. But since the exponential graph remains above the x-axis while the linear function becomes negative for $x < -20$, there must also be a second quadrant solution. We use different viewing window to see the two solutions.

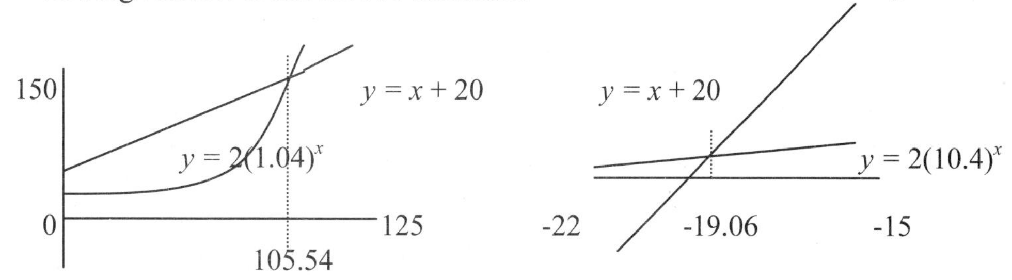

Answers: $t \approx 105.54$ or -19.06

2. $2\,P_0 = 1.12^t\,P_0$; $2 = 1.12^t$; $\ln 2 = \ln 1.12^t$; $\ln 2 = t \ln 1.12$; $t = \ln 2/\ln 1.12 \approx 6.12$ yrs.
3. $.88^t\,Q_0 = .5\,Q_0$; $.88^t = .5$; $\ln .88^t = \ln .5$; $t \ln .88 = \ln .5$; $t = \ln .5/\ln .88 \approx 5.42$ years
4. $\log A - \log B = 2$; $\log (A/B) = 2$; $A/B = 10^2 = 100$
5. $6.3\,(1.28)^t = 6.3\,e^{kt} = 6.3\,(e^k)^t$ if $1.28 = e^k$, so $\ln 1.28 = k \approx .247$. $Q = 6.3e^{0.247t}$

MASTERING CONCEPTS AND SKILLS

Use the √ , ?, * system on the exercises assigned by your instructor for this section.

Assigned Problems:

√

?

---

*

4.3 THE LOGARTHMIC FUNCTION

READING YOUR TEXTBOOK: Read Section 4.3, pp. 167-172.

As you read:

- Learn the domain of $\log x$. Recall that $\log x$ is only defined for positive numbers, because 10 to any power is always positive. (See example 3(f), p. 153.)
- Form a mental picture of the graph of $y = \log x$. The discussion on p.167 will help you understand the shape of the graph.
- Pay special attention to the relationship between the graphs of $y = 10^x$ and $y = \log x$, as shown in Figure 4.8

- Learn what it means to say that the line $x = 0$ (the y-axis) is a vertical asymptote of the graph of $y = \log x$. Say aloud what the notation "as $x \to 0^+$, $f(x) \to -\infty$" means.

- Form a mental picture of the graph of $y = \ln x$, and its relationship to the graph of $y = e^x$, as shown if Figure 4.10

- Examples 2-5 show some applications of logarithm functions in measuring quantities whose magnitudes vary widely. Note the use of properties of logarithms in Example 5.

REVIEWING THE BASICS

You should be able to:

- State the domain and range of the functions $y = \log x$ and $y = \ln x$.

- Identify vertical and horizontal asymptotes of $y = 10^x$, $y = \log x$, $y = e^x$, and $y = \ln x$.

- Sketch graphs of $y = 10^x$, $y = \log x$, $y = e^x$, and $y = \ln x$.

- State the relationship between the graphs of $y = \log x$ and $y = 10^x$, and between the graphs of $y = \ln x$ and $y = e^x$.

Practice Problems

1. Fill in the blanks. The domain of $y = \log x$ is ________________ and the range is ________________. The domain of $y = \ln x$ is ________________ and the range is ________________.

2. The notation "As $x \to 0^+$, $\ln x \to -\infty$" means ________________________ ________________________. Therefore the line $x = 0$ is a ______________ of the graph of $y = \ln x$.

3. Since $10^2 = 100$, the point (__ , __) is on the graph of $y = 10^x$, and the point (__ , __) is on the graph of $y = \log x$.

4. Circle the correct words. The graphs of $y = 10^x$ and $y = e^x$ are **increasing decreasing** and concave **up down**. The graphs of $y = \log x$ and $y = \ln x$ are **increasing decreasing** and concave **up down**.

5. Sketch the graphs of $y = e^x$ and $y = \ln x$ on the same coordinate system. Label all intercepts. Which graph has a vertical asymptote? ________ What is the equation of the

vertical asymptote? __________ Which graph has a horizontal asymptote?__________

What is the equation of the horizontal asymptote? _______________

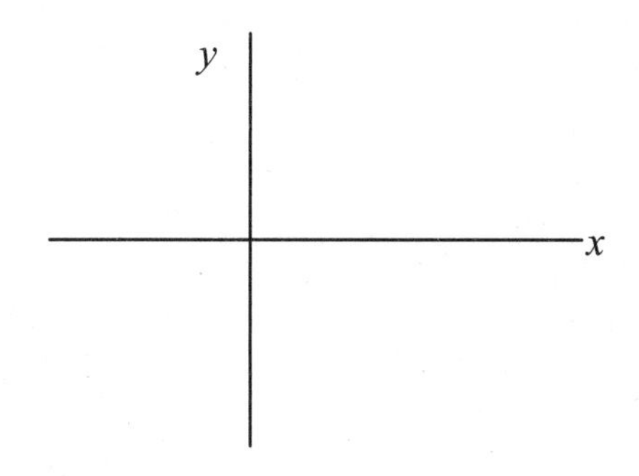

Solutions to Practice Problems

1. all positive numbers; all real numbers; all positive numbers; all real numbers.
2. As x gets closer to zero and is positive, ln x takes on larger and larger negative values; vertical asymptote
3. (2, 100); (100, 2)
4. increasing; up; increasing; down
5. $y = \ln x$; $x = 0$; $y = e^x$; $y = 0$

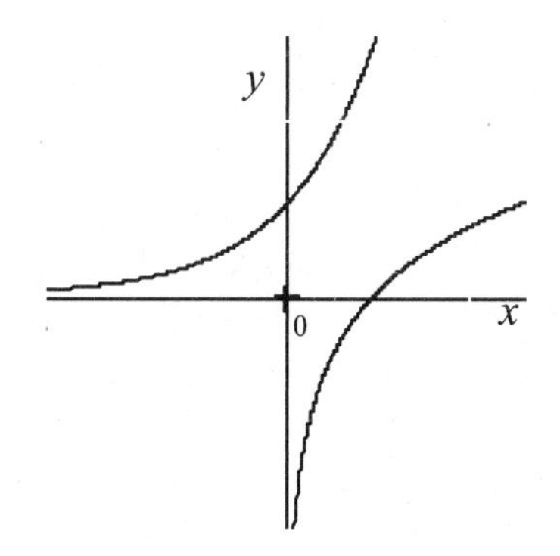

MASTERING CONCEPTS AND SKILLS

Use the √ , ?, * system on the exercises assigned by your instructor for this section.

Assigned Problems:

√

?

---

*

4.4 LOGARITHMIC SCALES

READING YOUR TEXTBOOK: Read Section 4.4, pp. 175-181.

As you read:

- Recognize "linear scale" as the way you generally mark the x and y axes in the plane.
- Notice that on a "logarithmic scale" equally spaced tick marks represent growth by the same multiplicative factor.
- Think about when you would want to use a logarithmic scale to represent numbers.
- Remember that the *difference* between equally spaced numbers can increase rapidly on a logarithmic scale.
- Notice that the lower left-hand corner on a log-scale is *not* (0,0).
- Be sure to do the calculations in the examples on your own calculator or computer.
- Notice footnote 10, p. 181. The numbers your calculator produces may be slightly different from the ones in the text.

REVIEWING THE BASICS

You should be able to:

- Compute the log of a given number.
- Compute x when $\log x$ is given.
- Decide when a log scale is more appropriate to use than a linear scale.
- Construct a log scale and use it to plot numbers.
- Transform an exponential equation, $N = a\,e^{kt}$, into a linear equation by taking the natural log of both sides, and using $y = \ln N$.
- Transform a linear equation into an exponential equation by exponentiating both sides.

Practice Problems

1. Compute the log of each of the following numbers:

   $a = 32100$ $\quad$ $b = 3.21$
   $g = 4.1\text{x}10^7$ $\quad$ $h = 4.1$
   $l = 2.89$ $\quad$ $m = .0289$
   $x = 0.147$ $\quad$ $y = 0.0000147$

2. (a) Write $10^7 - 10^6$ using standard notation.

   (b) Write $10^3 - 10^2$ using standard notation.

   (c) Compare the size of your answers.

3. (a) If $\log x = 5.4$, what is x?

   (b) If $\log y = 5.7$, what is y?

   (c) Compare your answers to (a) and (b).

4. For each of the following, decide whether a linear scale or a logarithmic scale would be more appropriate for plotting the given data.
   (a) The height, in inches, of members of your class.
   (b) The weight, in pounds, of all the animals in a large zoo.
   (c) The number of people watching each show on television at a given moment.
   (d) The number of words in each book in a city library.
   (e) The number of minutes of daylight each day of a given month where you live.

5. Use the number line drawn, and a log scale, to plot the following numbers:

   $A = 21$ $B = 389$ $C = 9000$ $D = 1900$ $E = 50{,}000$

   ____

   1 10 10^2 10^3 10^4 10^5

6. Transform the equation $P = 12.61\ e^{0.49\,t}$ into a linear equation in the form $y = mt + b$, where $y = \ln P$.

7. Transform the equation $\ln Q = 1.68\,t + 5.23$ into an exponential equation.

Solutions to Practice Problems

1. $\log a = 4.5065$ $\log b = 0.5065$
   $\log g = 7.6128$ $\log h = 0.6128$
   $\log l = 0.4609$ $\log m = -1.5391$
   $\log x = -0.8327$ $\log y = -4.8327$

2. (a) $10^7 - 10^6 = 10{,}000{,}000 - 1{,}000{,}000 = 9{,}000{,}000$
(b) $10^3 - 10^2 = 1000\text{-}100 = 900$
(c) $10^7 - 10^6$ is much larger then $10^3 - 10^2$.

3. (a) $x = 10^{5.4} \approx 251{,}189$
(b) $y = 10^{5.7} \approx 501{,}187$
(c) y is twice as big as x. A small change in the log can mean a large change in the value.

4. (a) linear scale. No one is even 10 times as tall as someone else.
(b) log scale. An elephant is many times larger than a small bird.
(c) log scale. A show seen nationwide will have many times the viewers as a local show on a small station.
(d) log scale. The number of words in a volume of an encyclopedia is many time the number of words in some children's books.
(e) linear scale. The number of minutes doesn't vary much in one location over a month.

5.

$A \quad B \quad D \quad C \quad E$

$1 \quad 10 \quad 10^2 \quad 10^3 \quad 10^4 \quad 10^5$

6. $\ln P = \ln(12.61\, e^{0.49\,t}) = \ln 12.61 + \ln e^{0.49\,t} = \ln 12.61 + 0.49\,t = 2.53 + 0.49\,t$, so the final answer is $y = 0.49\,t + 2.53$.

7. $e^{\ln Q} = e^{(1.68\,t + 5.23)} = e^{5.23}\ e^{1.68\,t} = 186.79\, e^{1.68\,t}$, so final answer is $Q = 186.79\, e^{1.68\,t}$.

MASTERING CONCEPTS AND SKILLS

Use the √ , ?, * system on the exercises assigned by your instructor for this section.

Assigned Problems:

√

?

*

CHAPTER FIVE

TRANSFORMATIONS OF FUNCTIONS AND THEIR GRAPHS

5.1 VERTICAL AND HORIZONTAL SHIFTS

READING YOUR TEXTBOOK: Read Section 5.1, pp. 194-199.

As you read:

- Study Example 1, p. 194, to see a physical interpretation of a vertical shift of a graph.
- Study Example 2, pp. 194-5 for an interpretation of a horizontal shift of the same graph.
- Study Examples 3 and 4 to see how these shifts affect the formula for the graph.
- Learn the summary in the box, p. 196. Try for understanding, and rely less on memorizing facts.
- Learn what is meant by a **translation** of a graph (p. 196).
- In applied problems, paying attention to units can help you distinguish between inside and outside changes. See Examples 5 and 6.
- Use your graphing calculator or computer to see the transformations in Example 9. First graph $y = x^2$. On the same set of axes, graph $y = (x - 2)^2$. Then graph $y = (x - 2)^2 - 1$. You will see these and other transformations of the graph of $y = x^2$ in Section 5.5 that deals with the family of quadratic functions. .

REVIEWING THE BASICS

You should be able to:

- Sketch the graphs of $y = g(x) + k$ and $y = g(x + k)$ if you know the graph of $y = g(x)$.
- Write the formula for a graph that is a translation of a given graph.

Practice Problems

Use the graph of the function f given below to answer Problems 1 and 2.

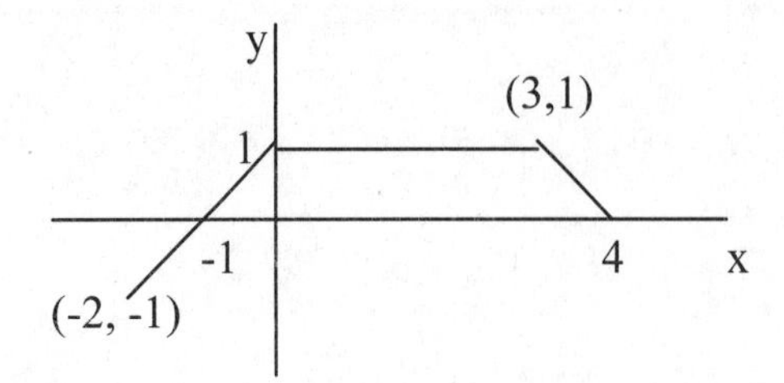

1. On the axes provided, sketch the graphs of:

(a) $y = f(x) + 2$

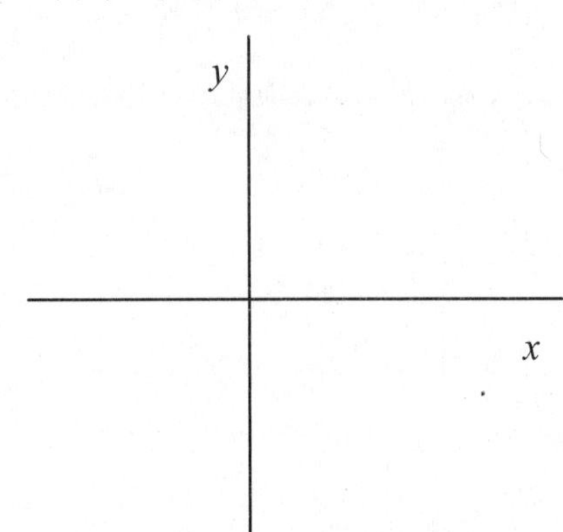

(b) $y = f(x + 2)$

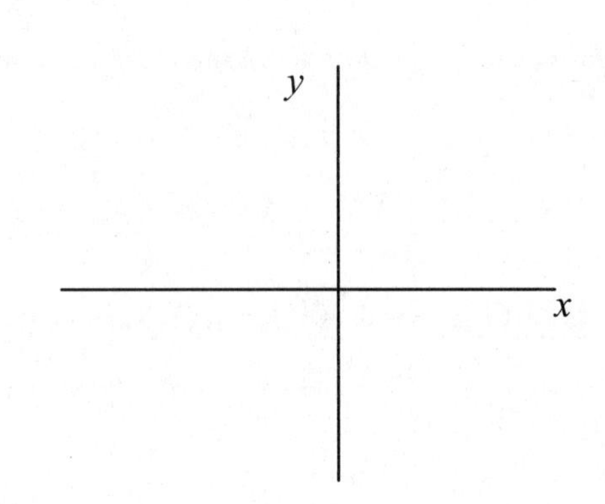

2. The graph of each function below is a translation of the graph of f. Write a formula, in terms of f, for each function.

(a)

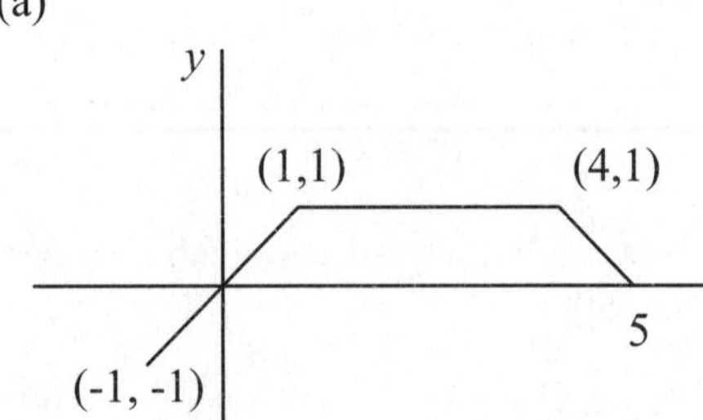

(b)

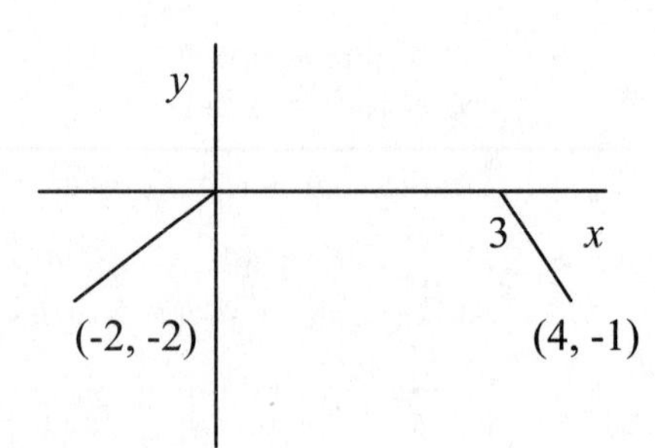

(c)

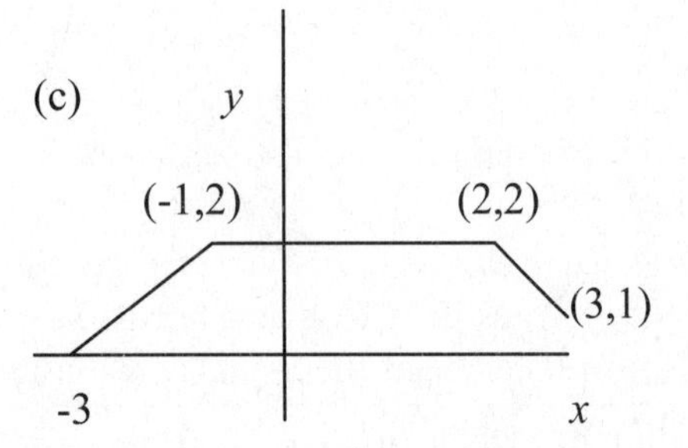

Solutions to Practice Problems

1. (a) (b)

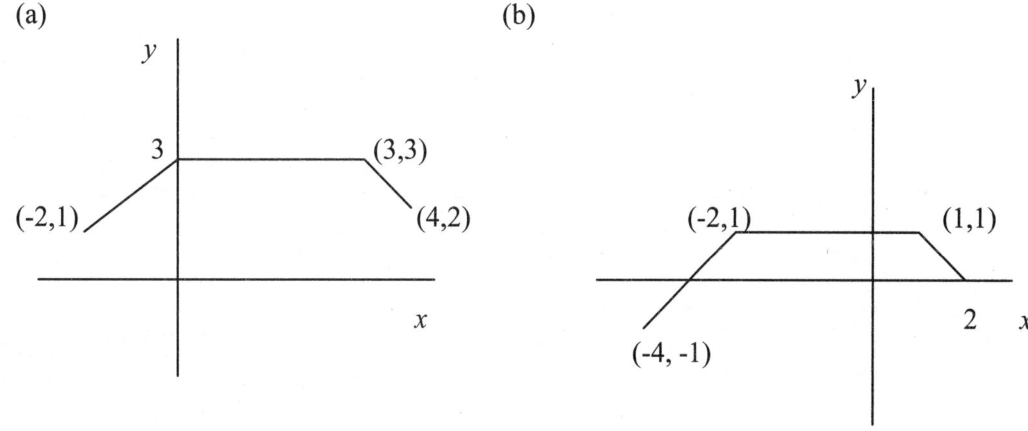

2. (a) $y = f(x - 1)$; (b) $y = f(x) - 1$; (c) $y = f(x + 1) + 1$

MASTERING CONCEPTS AND SKILLS

Use the √ , ?, * system on the exercises assigned by your instructor for this section.

Assigned Problems:

√

?

---

*

5.2 REFLECTIONS AND SYMMETRY

READING YOUR TEXTBOOK: Read Section 5.2, pp. 202-208.

As you read:

- Learn how the graphs of $y = -f(x)$ and $y = f(-x)$ are related to the graph of $y = f(x)$. (See pp. 202-207.) Know the summary (Box, p. 204).

- Notice especially in Figures 5.10-5.12 what happens to asymptotes when a graph is reflected. For example, the graph of f in Figure 5.9 has the x-axis as a horizontal asymptote. In Figure 5.10, the graph of $g(x) = -f(x)$ also has the x-axis as a horizontal asymptote. This happens because when $f(x)$ is near zero, so is $-f(x)$.

- Understand what it means to say that a graph is **symmetric about the *y*-axis**.
- Study Example 2, and learn the algebraic test to determine if the graph of a function is symmetric about the *y*-axis.
- Learn the definition of an **even function** (Box, p. 205). Form a mental picture of some even functions, such as $y = x^2$, $y = x^4$, and $y = |x|$.
- Learn what it means to say that a graph is **symmetric about the origin**. This type of symmetry is a little harder to visualize than symmetry about the *y*-axis, but Figure 5.15, p.206, should help.
- Learn the definition of an **odd function** (Box, p. 206). Form a mental picture of some odd functions, such as $y = x^3$ and $y = 1/x$.
- Study Example 5, p. 208, carefully. Notice especially why the line $y = 300$ is a horizontal asymptote, and what this means about the temperature of the yam.

REVIEWING THE BASICS

You should be able to:

- Sketch the graphs of $y = -f(x)$ and $y = f(-x)$ if you are given the graph of *f*.
- Determine algebraically if a function is odd, even, or neither.
- Determine by looking at its graph if a function is odd, even, or neither.
- Combine shifts and reflections of a given graph, and write a formula for the new graph.

Practice Problems

1. Fill in the blanks. The graph of $y = -f(x)$ is a reflection of the graph of $y = f(x)$ across the ____________________. A function *f* is called **even** if ______ = _________. The graph of an even function is symmetric about the ________________. A function *f* is called **odd** if _____ = _______. The graph of an odd function is symmetric about the ______________. If the line $y = b$ is a horizontal asymptote of the graph of $y = f(x)$, then the line $y =$ _________________ is a horizontal asymptote of the graph of $y = f(x) + k$.

2. The graph of a function f is given below.

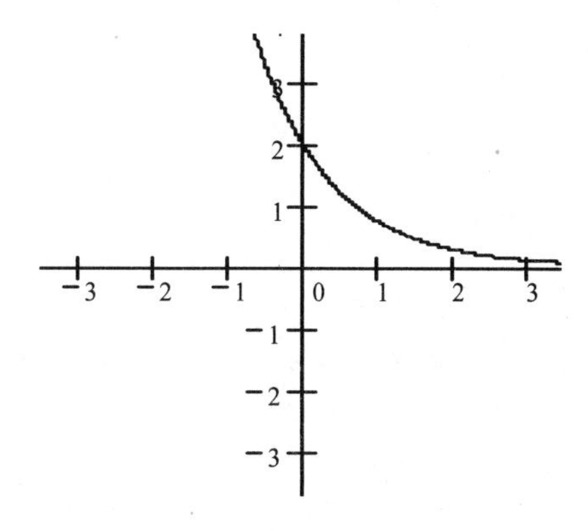

(a) Sketch the graph of $y = -f(x)$

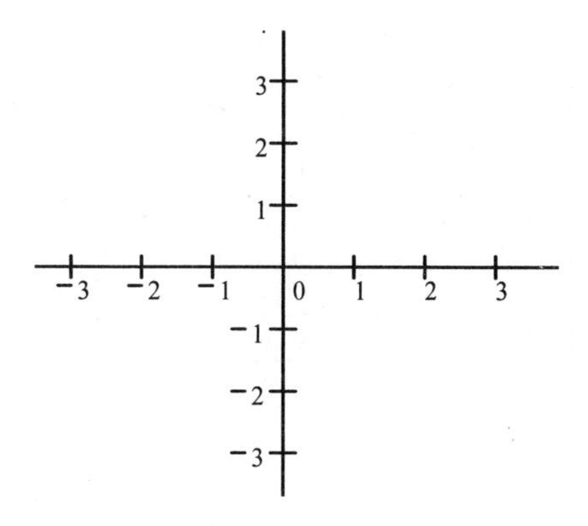

(b) Sketch the graph of $y = -f(x) + 3$. Indicate any asymptotes.

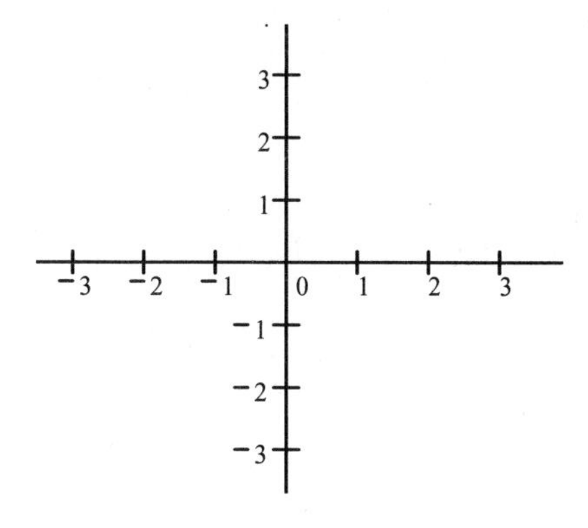

(c) Sketch the graph of $y = f(-x)$.

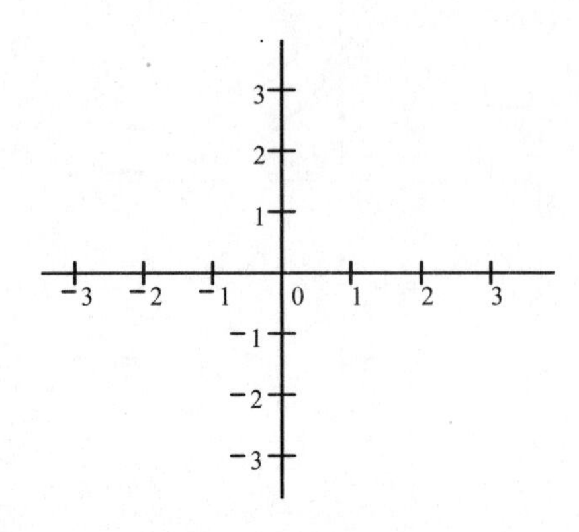

3. Determine algebraically whether each function is odd, even, or neither. Confirm your answer by graphing each function using your graphing calculator or computer.

   (a) $f(x) = |x| + 1$

   (b) $g(x) = |x + 1|$

   (c) $h(x) = 1/x^3$

4. The following graph is a transformation of the graph of $y = |x|$. Describe the transformations, and write a formula for the graph.

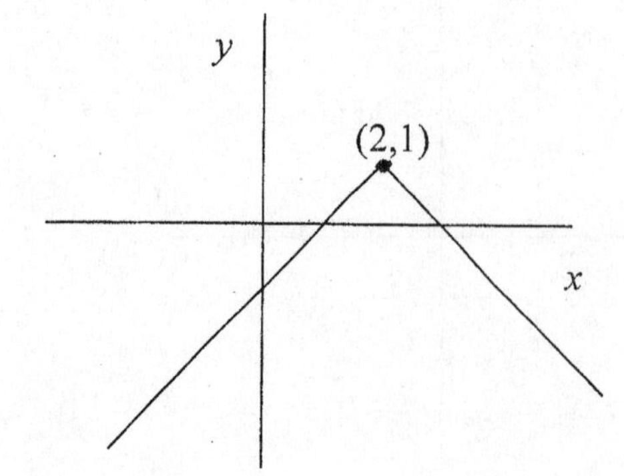

Solutions to Practice Problems

1. x-axis; $f(-x) = f(x)$; y-axis; $f(-x) = -f(x)$; origin; $y = b + k$.

2. (a)

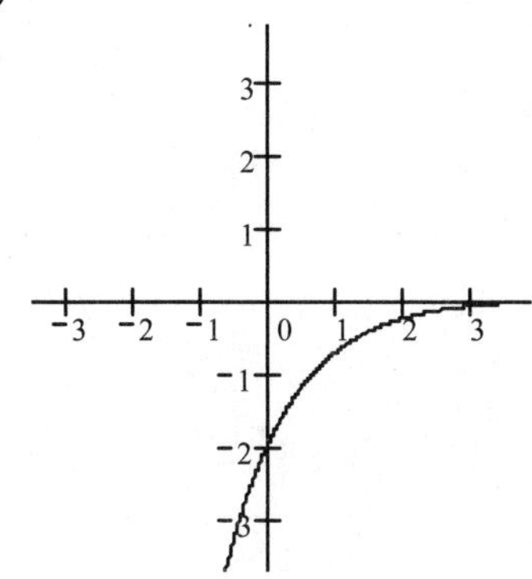

(b)

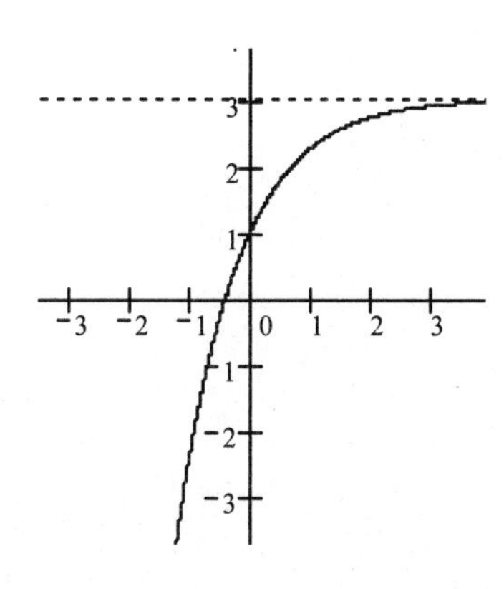

(c)

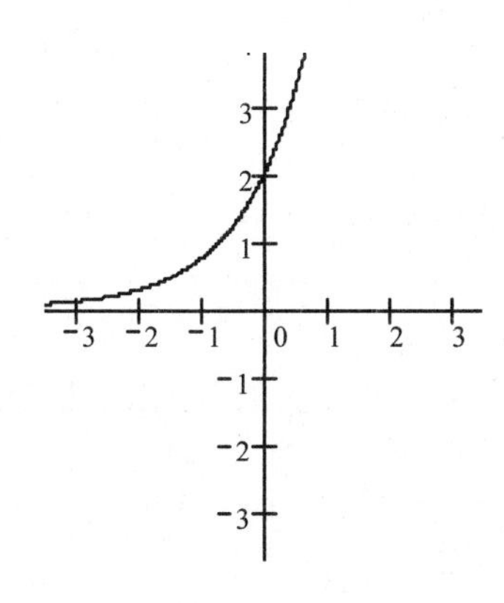

3. (a) $f(-x) = |-x| + 1 = |x| + 1 = f(x)$; so f is even.

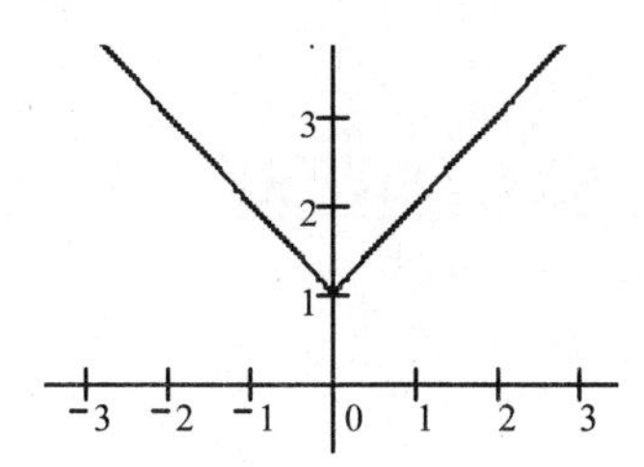

(b) $g(1) = 2$, but $g(-1) = 0$, so g is neither odd nor even.

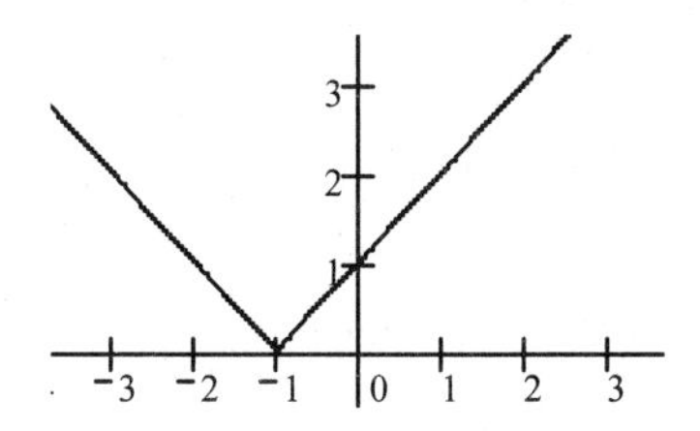

(c) $h(-x) = \frac{1}{(-x)^3} = \frac{1}{-x^3} = -\frac{1}{x^3} = -h(x)$; so h is odd.

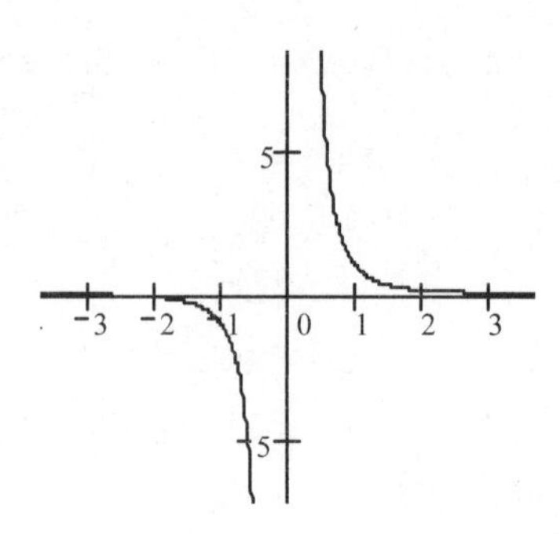

4. The graph of $y = |x|$ has been (i) shifted right two units, (ii) reflected about the x-axis, and (iii) shifted up one unit. The formula is $y = -|x-2| + 1$.

MASTERING CONCEPTS AND SKILLS

Use the √ , ?, * system on the exercises assigned by your instructor for this section.

Assigned Problems:

√

?

---

*

5.3 VERTICAL STRETCHES AND COMPRESSIONS

READING YOUR TEXTBOOK: Read Section 5.3, pp. 211-216.

As you read:

- Learn what is meant by a **vertical stretch** of a graph, and why the horizontal intercepts are not changed by a vertical stretch (p. 211-212).

- *Understand* the summary of vertical stretch or compression (Box, p. 212). If you understand Figure 5.25, you will understand the summary, and will not need to rely on memory.

- Follow all the calculations in Example 2, pp. 214-215, and in the generalization that follows (p. 215) which leads to the general effect of a vertical stretch on average rates of change (Box, p. 215).

- Follow the 4 steps of the sequence of transformations in Example 3, pp. 215-216. You will see this type of sequence of transformations again in Section 5.5.

REVIEWING THE BASICS

You should be able to:

- Sketch the graph of $y = k{\cdot}f(x)$ if you are given the graph of $y = f(x)$.

- Sketch the graph of a function obtained by a sequence of transformations of a given function.

- Write a formula for a graph that is a transformation of a given graph.

Practice Problems

Use the graph of the function *f* given below to answer Problems 1-3.

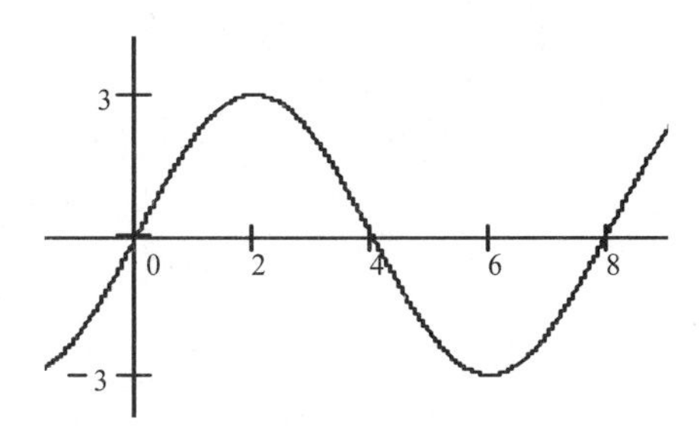

1. Sketch the graph of $g(x) = 2{\cdot}f(x)$

2. Sketch the graph of $h(x) = -\frac{1}{2} \cdot f(x)$

3. Find a formula, in terms of f, of the function whose graph is shown below.

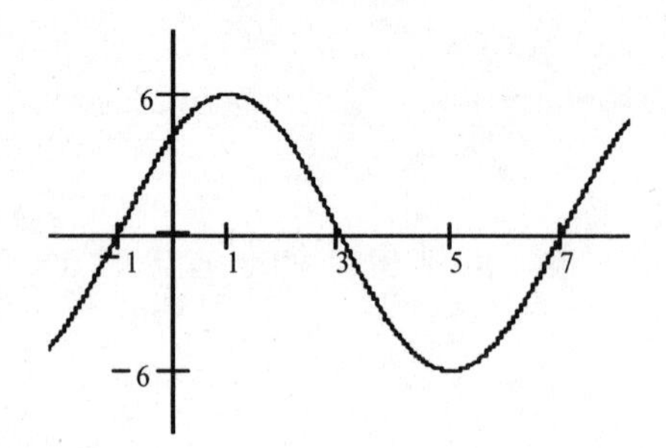

<u>Solutions to Practice Problems</u>

1.

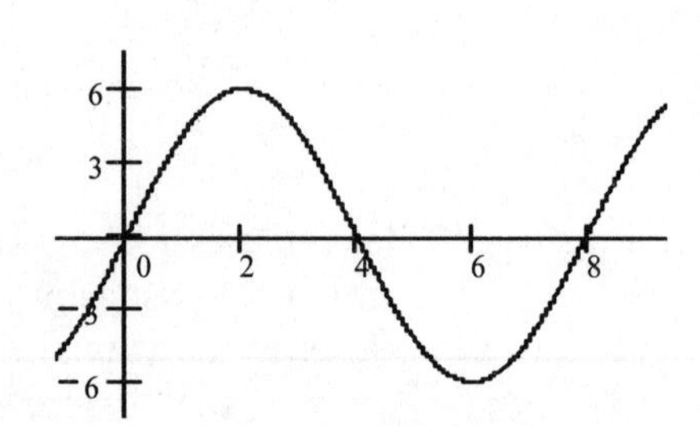

2.

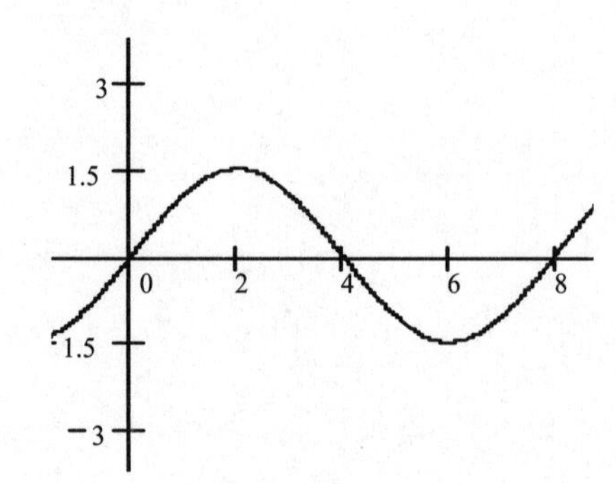

3. $y = 2 \cdot f(x + 1)$

MASTERING CONCEPTS AND SKILLS

Use the √ , ?, * system on the exercises assigned by your instructor for this section.

Assigned Problems:

√

?

---

*

5.4 HORIZONTAL STRETCHES AND COMPRESSIONS

READING YOUR TEXTBOOK: Read Section 5.4, pp. 219-222.

As you read:

- Study the discussion of the lighthouse beacon, pp. 219-220, to see a physical example of horizontally stretching or compressing a graph. Then learn the generalization (Box, p. 221).

- Follow the use of function notation carefully in Examples 1 and 2, pp.221-222, so that you understand the tables and graphs for the new functions g and h.

- Notice in Example 3 that you are looking at horizontal compression and stretching of the graph of $f(t) = e^t$. So the new functions have the form $y = e^{kt}$. Just as in Examples 1 and 2, the larger the value of k, the more compressed (closer to the y-axis) the graph. This result is consistent with the way you compared exponential graphs in Chapter 3. Rewriting each function as $y = (e^k)^t$, you get the steepest graph when the base is largest—that is, when the value of k is largest.

REVIEWING THE BASICS

You should be able to:

- Sketch the graph of $g(x) = f(kx)$ if you know the graph of f.

- Recognize a graph as a horizontal stretching or compression of another graph.

Practice Problems

Use the following graph of the function f to answer Problems 1-3.

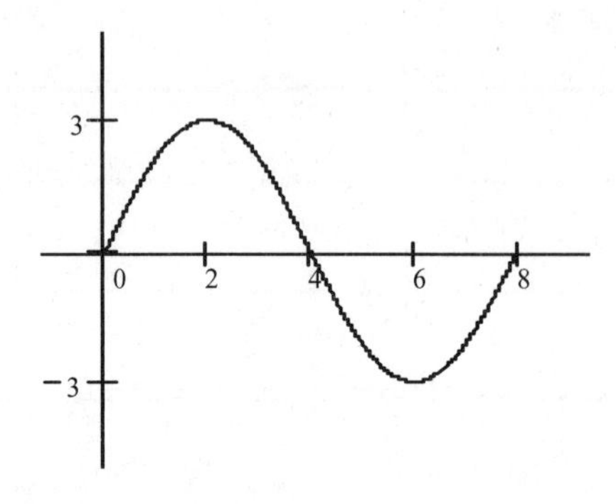

1. Sketch the graph of $y = f(2x)$.

2. Sketch the graph of $y = f(.5x)$.

3. Find a formula, in terms of f, for the following graph.

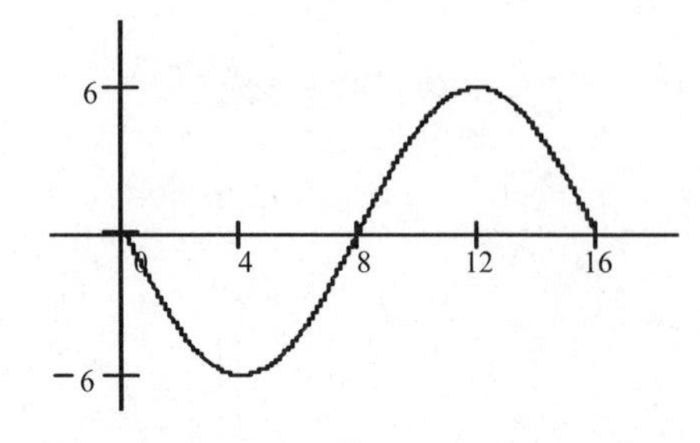

Solutions to Practice Problems

1.

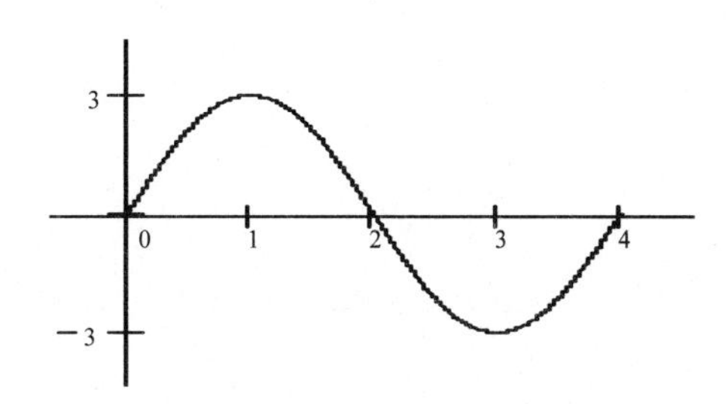

2.

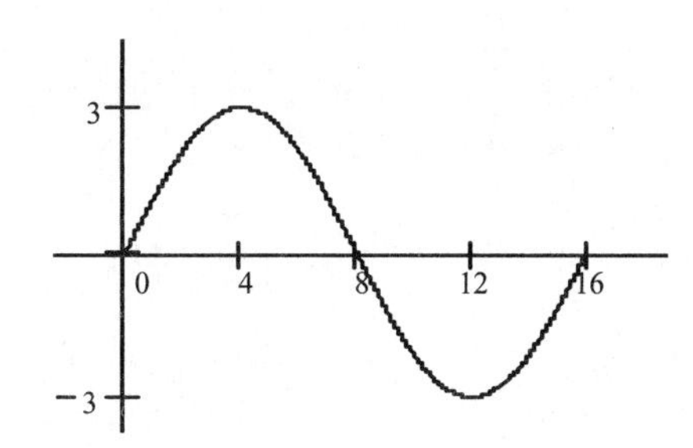

3. $y = -2f(.5x)$

MASTERING CONCEPTS AND SKILLS

Use the √, ?, * system on the exercises assigned by your instructor for this section.

Assigned Problems:

√

?

---

*

5.5 THE FAMILY OF QUADRATIC FUNCTIONS

READING YOUR TEXTBOOK: Read Section 5.5, pp. 225-230.

As you read:

- Notice that in this section you are looking at transformations of the function $f(x) = x^2$.
- Learn the general form of a **quadratic function**, and form a mental picture of a **parabola**.
- Follow the transformations of the graph of f that result in the graph of g in Example 1, pp. 225-226.
- Read the generalization in the middle of p.226. Notice that the graph is a translation of the graph of $y = ax^2$. So if $a < 0$, the graph opens downward.
- Visualize what are meant by the **vertex** and the **axis of symmetry** of a parabola.
- Know the summary and all the terms in the box on p. 227. If you think in terms of transformations of the graph of $y = x^2$, you will understand the **vertex** form, and you can rely less on memorizing facts.
- Review **Completing the Square** in the Tools Section, pp. 239-240. You need to know this to change a quadratic function from standard form to vertex form.
- Follow the algebra in Example 2, pp. 227-228. Be especially careful factoring out –4 in part (b).
- Notice that in Examples 3 and 4 you are asked to **find a formula for a given parabola,** but how you proceed depends on the information given. In Example 3, the vertex is given, so you know h and k in the vertex form. Then use the coordinates of the other given point to find a. In Example 4, you do not know the vertex, but you do know three points. In this case, use the factored form. Work carefully through the arithmetic in both examples.
- Notice that in the standard form, **the parameter c is the y-intercept,** or $f(0)$.
- **Finding the x-intercepts** (zeros) requires solving the equation $f(x) = 0$. Notice the difference between this and the previous bullet.
- Study Example 5 carefully. Look back at Figure 5.52, p.225. Remember that the y-coordinate of a point on the graph gives the height of the ball. The largest y-value is at the vertex, so the y-coordinate of the vertex gives the maximum height of the ball. Follow the steps in completing the square in Example 5 to find the vertex.

- Notice the wording of the problems in Example 6. We are asked to find the maximum area. To answer the question, we first have to write a function that gives the area. Make sure you understand how the length of fencing is used to get the area function and how you use the vertex to get the maximum area.

REVIEWING THE BASICS

You should be able to:

- Change a quadratic function from standard form to vertex form by completing the square.
- Sketch (without a calculator or computer) the graph of a quadratic function given in vertex form.
- Find a formula for a parabola if you know the vertex and one other point.
- Find a formula for a parabola if you know the *x*-intercepts and one other point.
- Find a formula for a parabola if you know the *y*-intercept and two other points.
- Find the zeros (*x*-intercepts) of a quadratic function by factoring or using the quadratic formula.
- Find maximum or minimum values of a given quadratic function.

Practice Problems

1. Let $f(x) = -\frac{1}{2}(x+2)^2 - 1$. Describe a sequence of transformations of the graph of $y = x^2$ which result in the graph of *f*, and sketch the graph of *f*. Label the vertex and the *y*-intercept.

2. Let $f(x) = x^2 - 4x - 5$.

   (a) Change $f(x)$ to vertex form by completing the square, and identify the vertex.

   (b) Find the x- and y-intercepts.

   (c) Sketch the graph. Label the vertex and all intercepts.

3. Find a possible formula for each parabola.

   (a)

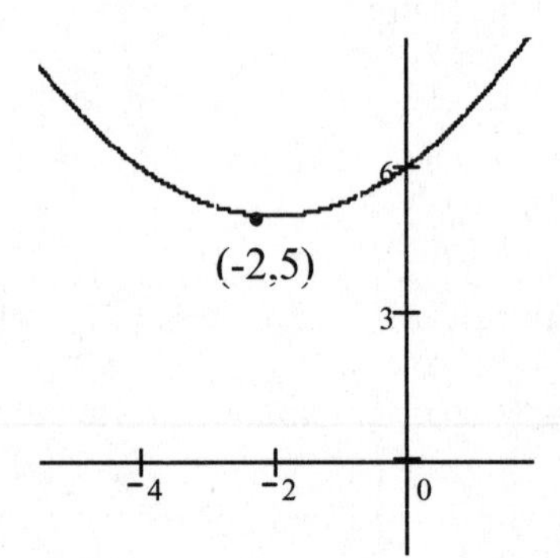

   (b)

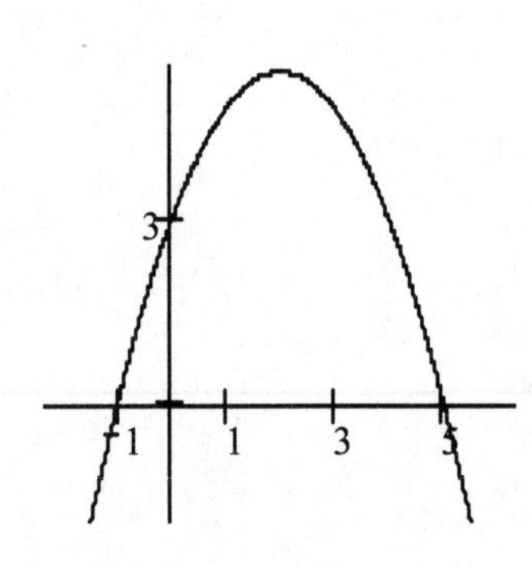

4. Try to use the quadratic formula to find the zeros of $f(x) = x^2 - 4x + 5$. What do you conclude about the graph of f?

5. Find the minimum value of $f(x) = x^2 - 6x + 5$. Explain your reasoning.

Solutions to Practice Problems

1. The graph of $y = x^2$ is (i) shifted left two units, (ii) shrunk vertically by a factor of 1/2, (iii) reflected about the x-axis, and (iv) shifted down one unit. y-intercept: $f(0) = -3$

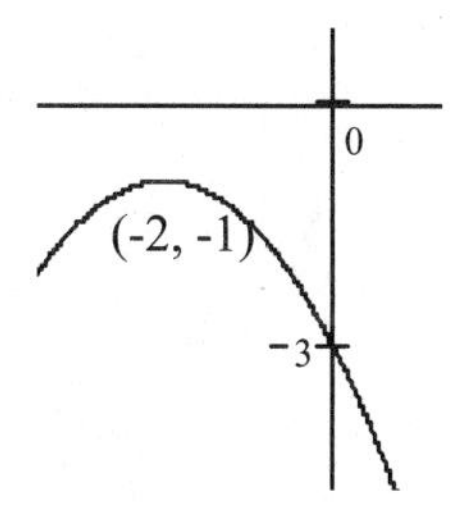

2. (a) $f(x) = (x^2 - 4x + 4 - 4) - 5 = (x^2 - 4x + 4) - 4 - 5 = (x - 2)^2 - 9$. Vertex $(2, -9)$
   (b) $f(0) = -5$ is the y-intercept.
   x-intercepts: $x^2 - 4x - 5 = (x - 5)(x + 1) = 0$ when $x = 5$ or $x = -1$.
   (c)

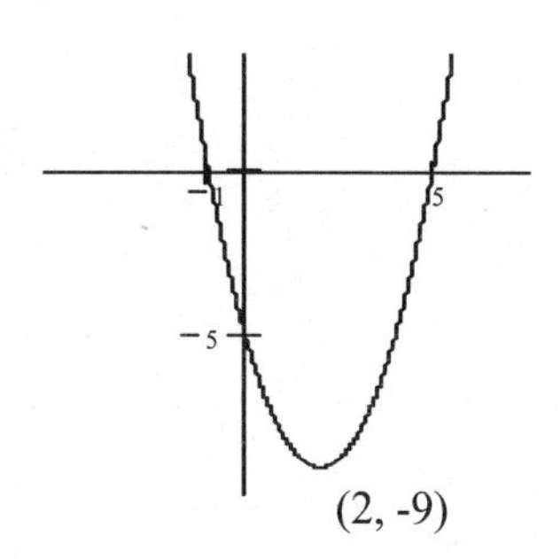

3. (*a*) $y = a(x + 2)^2 + 5$; $y = 6$ when $x = 0$, so $6 = a(2)^2 + 5$; $1 = 4a$, or $a = .25$. The equation is $y = .25(x + 2)^2 + 5$.
   (b) $y = a(x + 1)(x - 5)$; $y = 3$ when $x = 0$, so $3 = a(1)(-5) = -5a$; $a = -.6$. The equation is $y = -.6(x + 1)(x - 5)$

4. $x = (4 \pm \sqrt{16-4(5)})/2 = (4 \pm \sqrt{-4})/2$, which is not a real number. The graph of f has no x-intercepts.
5. The graph of f is a parabola opening up, so the minimum y-value is at the vertex. The vertex form is $f(x) = x^2 - 6x + 9 - 9 + 5 = (x-3)^2 - 4$ So the vertex is (3, –4), and the minimum value of f is –4.

MASTERING CONCEPTS AND SKILLS

Use the √ , ?, * system on the exercises assigned by your instructor for this section.

Assigned Problems:

√

?

---

*

CHAPTER SIX

TRIGONOMETRIC FUNCTIONS

6.1 INTRODUCTION TO PERIODIC FUNCTIONS

READING YOUR TEXTBOOK: Read Section 6.1, pp. 244-247.

As you read:

- Keep in mind that the function being discussed gives your height above the ground t minutes after boarding the ferris wheel and that the ferris wheel takes 30 minutes to complete one rotation. Be sure you understand the entries in Table 6.1, p. 245.

- Study the top of p. 246 to see why the dots in Figure 6.4, p. 245, are not joined by line segments. Try to understand why the graph is wave-shaped and why the curve repeats itself.

- Learn what it means to say that a function is **periodic with period c** (Box, p.247).

- Learn what the **amplitude** and **midline** of a periodic function tell you about the graph (Box, p. 247).

REVIEWING THE BASICS

You should be able to:

- Sketch a graph of your height t minutes after boarding a ferris wheel if you know the diameter of the wheel, the time it takes for one complete rotation, and your boarding position. Label the period, midline, and amplitude of the graph.

- Decide whether or not a function is periodic by looking at the graph.

- Determine the midline, amplitude, and period of a wave-shaped graph, and interpret these quantities physically.

Practice Problems

1. Sketch the graph of your height in feet above the ground t minutes after boarding a ferris wheel that is 200 feet in diameter and makes one complete rotation every 8 minutes. Assume you board at the six o'clock position. Label the amplitude, period, and midline.

2. Determine the midline, amplitude, and period of the following graph.

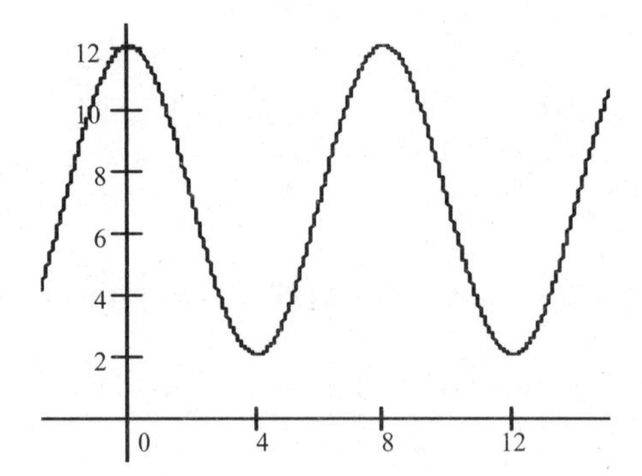

Solutions to Practice Problems

1.

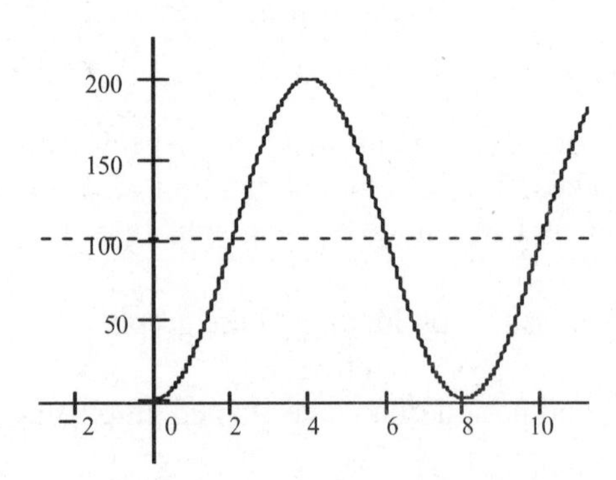

2. midline y = 7; amplitude 5; period 8

MASTERING CONCEPTS AND SKILLS

Use the √ , ?, * system on the exercises assigned by your instructor for this section.

Assigned Problems:

√

?

---

*

6.2 THE SINE AND COSINE FUNCTIONS

READING YOUR TEXTBOOK: Read Section 6.2, pp. 250-255.

As you read:

- Learn the equation of the **unit circle**. Have a mental picture of the unit circle and the points (1,0), (0,1), (-1,0), and (0,-1).
- Learn how to **associate a point on the unit circle with a given angle θ.** Notice that the point is on the terminal side of θ, and is 1 unit from the origin (Figure 6.13).
- Learn the **definition of sine θ and cosine θ** (Box, p. 251 and Figure 6.17). This definition is the key to understanding all of Chapter 6. When you understand this definition, you will not need to rely on a lot of memorized facts.
- Be sure you see how the coordinates labeled in Figures 6.14-6.16, p. 251, give the answers in Example 2, p. 252.
- Notice that both coordinates of the point R in Figure 6.16, p. 251, are negative.
- Notice in Example 3 that the angle is measured *from the positive x-axis*. It is not an angle in the triangle formed by the radius and the lengths m and n. Understand how the lengths m and n, are related to the sine and cosine of the angle (which are the coordinates of point Q). The *length* of line segment n is a positive number, but the x-coordinate of the point Q is negative.
- Study the argument using similar triangles that gives the very important formulas for the coordinates of point Q in Figure 6.18; see box p.252. These formulas are used in Example 4. Notice that to find the coordinates of the point C, we use an angle *measured from the positive x-axis* which has point C on the terminal side.

- Notice in Example 6 that the formula $225 + y$ for your height above the ground is valid no matter where you are. When you are on the bottom half of the ferris wheel the *y*-coordinate will be negative, so the formula gives a height less than 225, as desired.

- Study the explanation and sketches, p. 254 of the conventional way of measuring angles.

REVIEWING THE BASICS

You should be able to:

- Associate a point on the unit circle with a given angle θ, and define the sine and cosine of θ in terms of the coordinates of that point.

- Use trig functions and your calculator to find the coordinates of a point P on the unit circle associated with a given angle θ.

- Use trig functions and your calculator to find the coordinates of the point P associated with an angle θ on a circle of any radius.

Practice Problems

Find the coordinates of each of the points.

1.

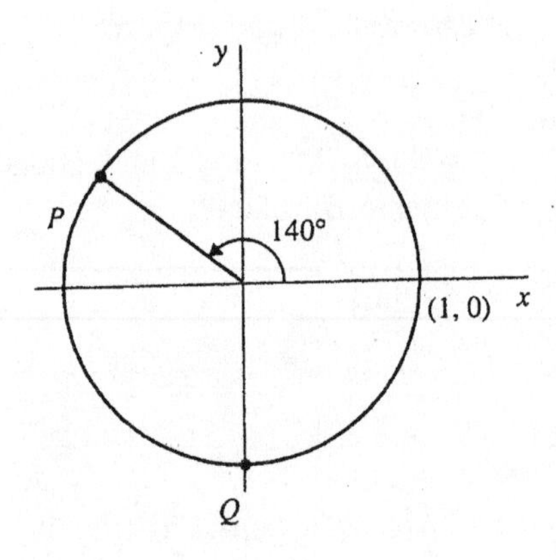

P: $x =$ __________ $\approx$ __________ $y =$ __________ $\approx$ __________

Q: $x =$ ______ $y =$ ______

2.

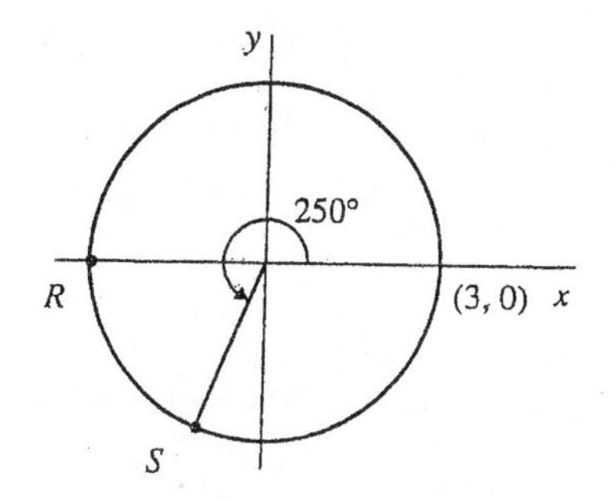

R: $x =$ ______ $y =$ ______

S: $x =$ __________ $\approx$ __________ $y =$ ______________ $\approx$ ____________

Solutions to Practice Problems

1. P: $x = \cos 140^\circ \approx -0.766$, $y = \sin 140^\circ \approx 0.643$; Q: $x = 0, y = -1$
2. R: $x = -3, y = 0$; S: $x = 3 \cos 250^\circ \approx -1.026$; $y = 3 \sin 250^\circ \approx -2.819$

MASTERING CONCEPTS AND SKILLS

Use the √ , ?, * system on the exercises assigned by your instructor for this section.

Assigned Problems:

√

?

---

*

6.3 RADIANS

READING YOUR TEXTBOOK: Read Section 6.3, pp. 257-261.

As you read:

- Learn the definition of **radian measure of an angle** (Boxes, pp. 257 and 258, and Figures 6.29 and 6.30).
- Develop a feeling for the position on the unit circle of the points corresponding to 1 radian, 2 radians, 3 radians, and so on. (See Figure 6.31 and Example 1, p. 258.) It might help to cut a small piece of string the length of the radius of the circle in

Figure 6.31, and use the string to measure off arcs, starting at the point (1,0), of lengths 1 radius, 2 radii, 3 radii, and so on. It also helps to keep in mind that a complete revolution spans the circumference, which for the unit circle is $2\pi \approx 6.28$. So the radian measure of a complete revolution is $2\pi \approx 6.28$ radians, of a half-revolution is $\pi \approx 3.14$ radians, and of a right angle, or a quarter-revolution, is $\pi/2 \approx 1.57$ radians. Understand how these key angles help you arrive at the answers to Example 1.

- Notice that when you convert between degrees and radians (Example 2, p. 258), you only need to use one fact: 2π radians = 360°. You can get both conversion factors from this one fact.

- Remember when you use the arc length formula (Box, p. 260) that the angle must be measured in radians. See Example 4, p. 260.

- It may help to remember that the radian measure of an angle is a way of measuring the angle in terms of the length of the radius. The radian measure of an angle is the number of radial lengths in the arc spanned by the angle (Figure 6.34). That is, in radian measure, $\theta = s/r$, where s is the length of the arc spanned by θ (Box, p. 260).

- Recall Figure 6.34, as you study Example 6, p. 261. Notice that on the circle with radius 1 mile, an arc of length 4 miles spans an angle of 4 radians; but on a circle with radius 3 miles, an arc of length 4 miles spans an angle of 4/3 radians.

REVIEWING THE BASICS

You should be able to:

- Define **radian measure of an angle**, and illustrate with a sketch (as in Figures 6.29 and 6.30).

- Locate (approximately) points on the unit circle corresponding to angles with given radian measures.

- Convert between radian measure and degree measure of an angle.

- Explain the radian measure of an angle in terms of the radius and the length of the arc spanned by the angle.

- Use the relationship $s = r\theta$ in applied problems.

Practice Problems

1. On the unit circle given below, label the points corresponding to angles having the given radian measure.

   A: 1.5 radians B: 3 radians C: -2 radians

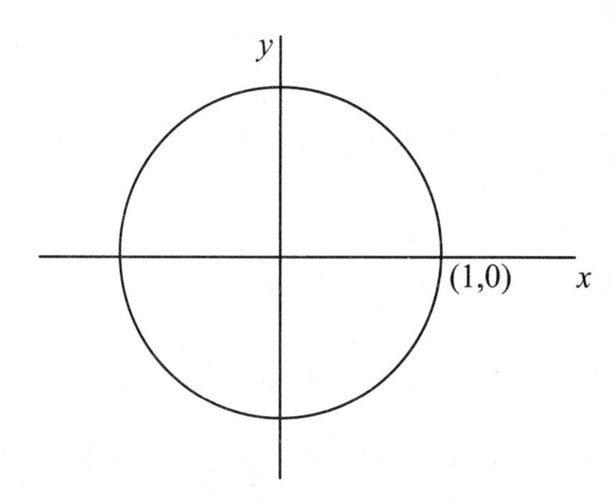

2. Determine the radian measure of the angle θ in each sketch

(a)

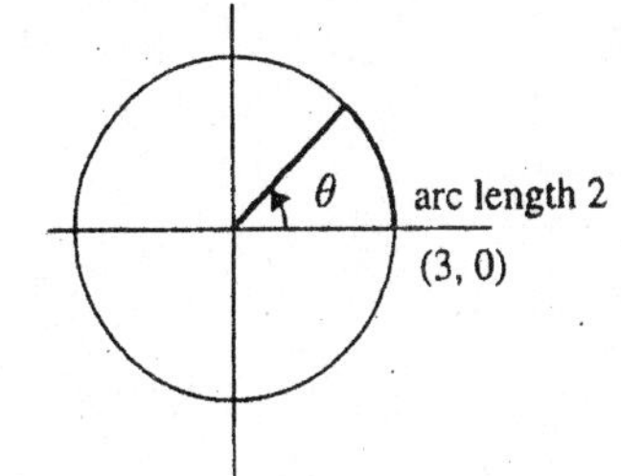

(b)

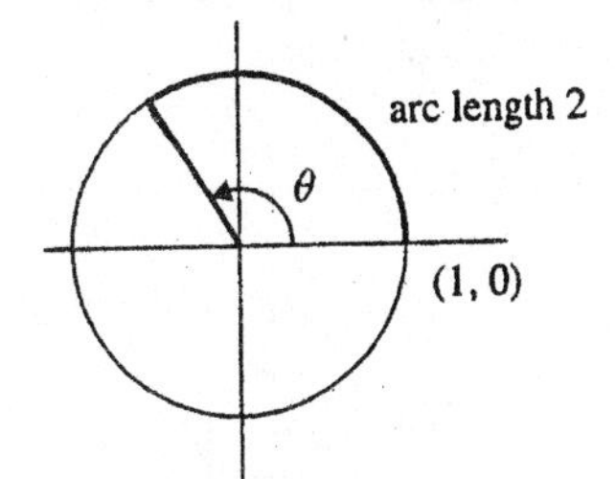

3. Find the length of the arc s.

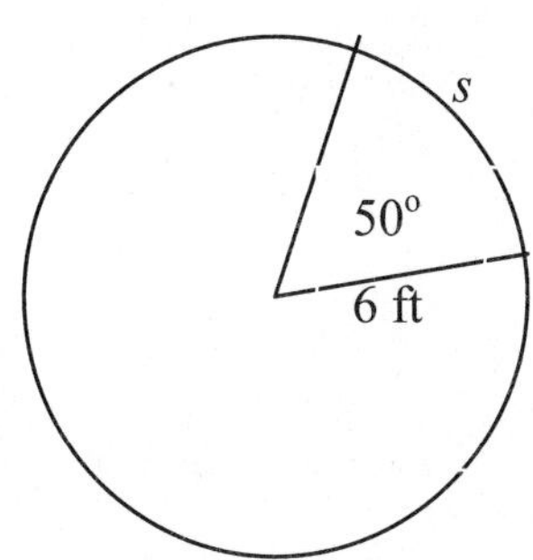

Solutions to Practice Problems

1.

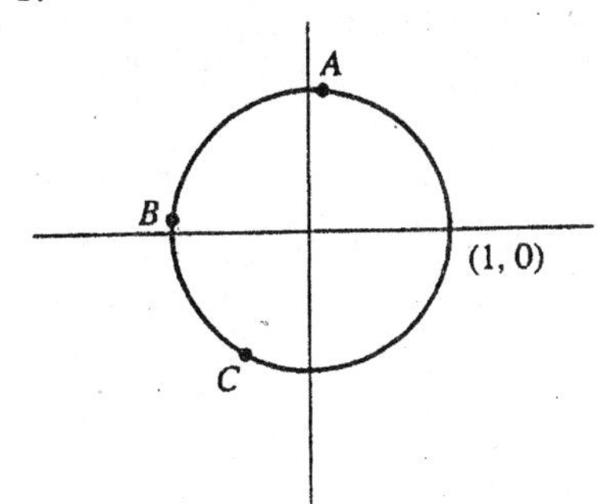

2. (a) θ = arc length / radius, so $\theta = 2/3$ radian (b) $\theta = 2$ radians

3. $\theta = 50^{\circ} \cdot \frac{\pi}{180^{0}} = \frac{5}{18}\pi$; $s = 6 \cdot \frac{5}{18}\pi \approx 5.24$ ft

MASTERING CONCEPTS AND SKILLS

Use the √ , ?, * system on the exercises assigned by your instructor for this section.

Assigned Problems:

√

?

---

*

6.4 GRAPHS OF THE SINE AND COSINE

READING YOUR TEXTBOOK: Read Section 6.4, pp. 263-266.

As you read:

- Keep in mind the definition of the sine and cosine (Box, p. 251) as you read pp. 263-264.
- Notice that some of the entries in Table 6.5, p. 264, are approximations of the exact values.
- Form a mental picture of the graphs of sine and cosine, with radians as units on the θ-axis. You should know where each graph crosses the θ-axis and the coordinates of the turning points of each graph. Notice that the θ-intercepts of one graph correspond to turning points of the other graph. Also notice the range and the period of each function.
- Recall what you learned in Chapter 5 about how the graphs of $y = k \cdot f(x)$ and $y = f(x) + b$ are related to the graph of $y = f(x)$. Relate this to graphs of sine and cosine in Example 2 and in the generalization following the example. Use both types of transformations in the ferris wheel function in Example 3.

REVIEWING THE BASICS

You should be able to:

- Sketch graphs of $y = \sin x$ and $y = \cos x$. Label intercepts and x-coordinates of turning points.
- State the domain, range, and period of the sine and cosine functions.
- Sketch the graph of a function of the form $y = A \sin x + B$ or $y = A \cos x + B$, and identify the amplitude and the midline.

Practice Problems

1. Sketch the graph of $y = \cos x$ on the axis below. Label the x-intercepts and the coordinates of all turning points.

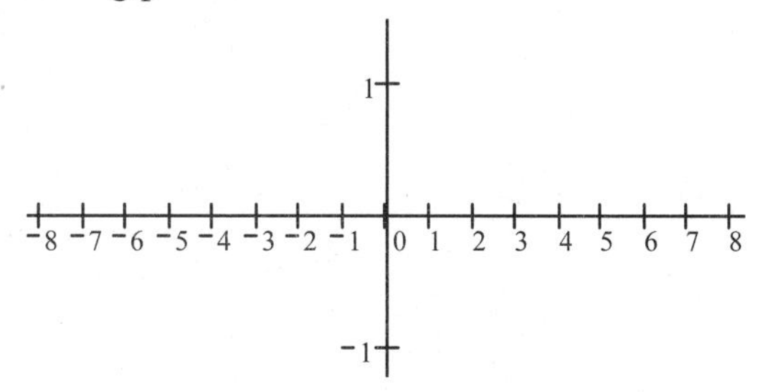

2. Fill in the blanks. The domain of both the sine and cosine functions is ____________, and the range of both functions is ______________. Both functions have period ___, amplitude ___, and midline ___.

3. Match each graph in parts (a) – (d) with the corresponding formula from (i) – (iv). You should be able to do this without a graphing calculator or computer.

   (i) $y = -3 \cos x$ (ii) $y = -3 + \cos x$ (iii) $y = 2 \sin x + 1$ (iv) $y = -3 \sin x$

   (a)

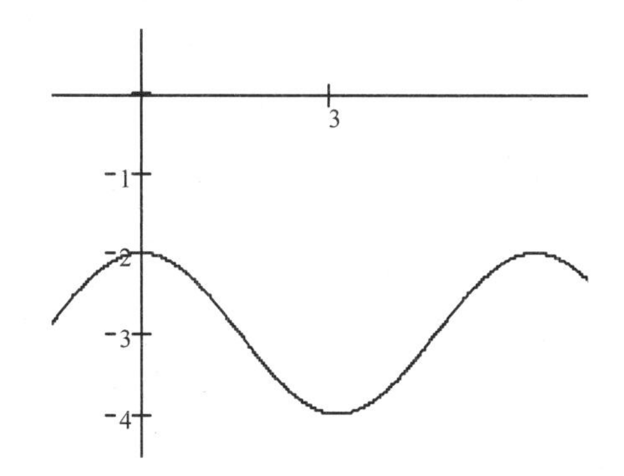

   (b)

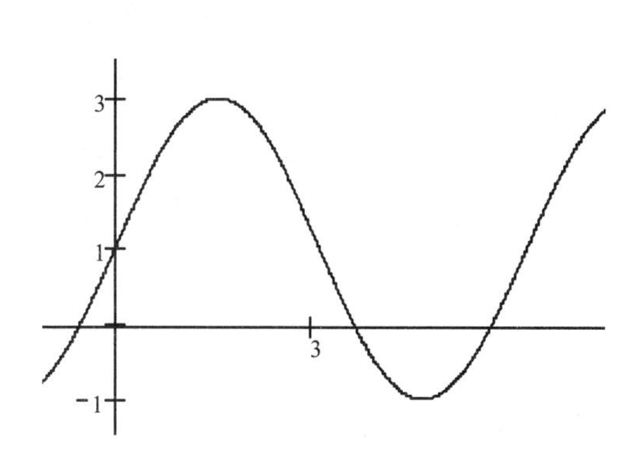

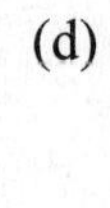

(c)

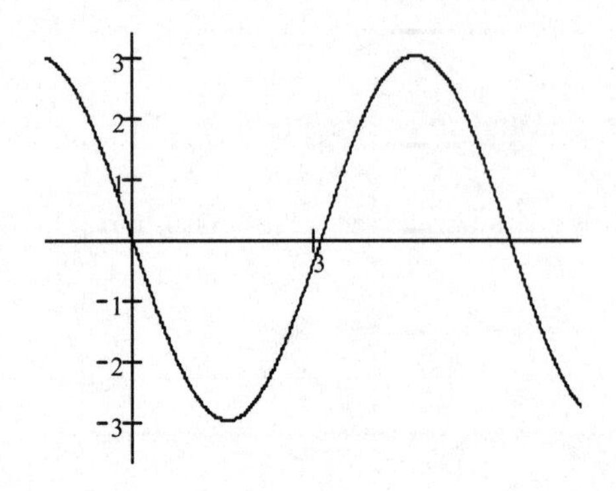

(d)

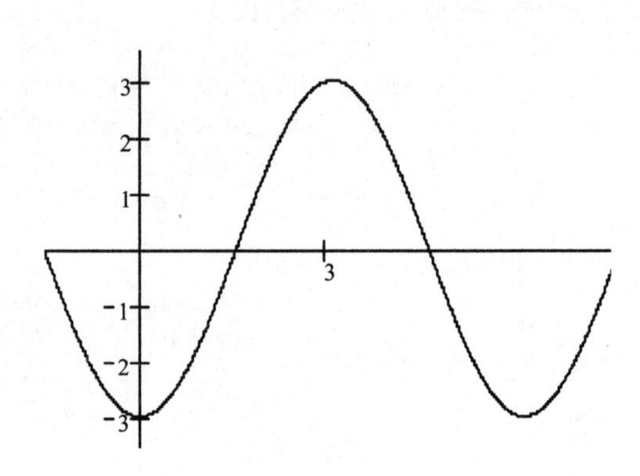

Solutions to Practice Problems

1.

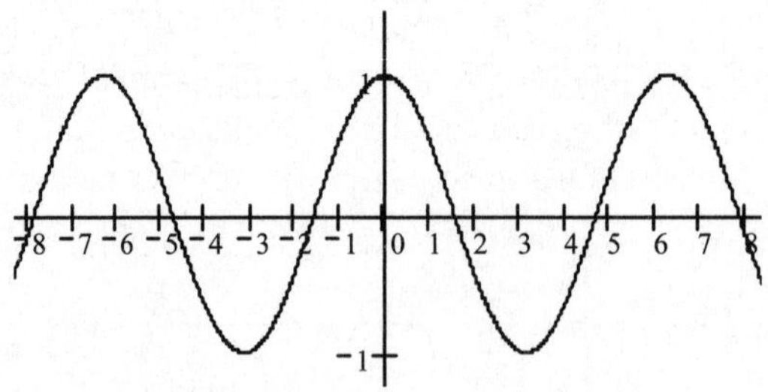

2. all real numbers; [-1,1]; 2π; 1; $y = 0$
3. (a) ii; (b) iii; (c) iv; (d) i

MASTERING CONCEPTS AND SKILLS

Use the √ , ?, * system on the exercises assigned by your instructor for this section.

Assigned Problems:

√

?

---

*

6.5 SINUSOIDAL FUNCTIONS

READING YOUR TEXTBOOK: Read Section 6.5, pp. 269-275.

As you read:

- Learn what is meant by the term **sinusoidal** function (p. 269). You have already learned the effect of the parameters *A* and *k* (Section 6.4). In this section, you will consider the effect of the other two parameters *B* and *h*.

- Recall that you learned in Chapter 5 about horizontal compression and stretching. Example 1 shows you how horizontal stretching and compression affect the period of a sinusoidal function.

- Notice that the parameter *B* is not the period. You should memorize the relationship between the period *P* and the parameter *B* (Box, p. 270).

- Use that relationship to get the value of *B* when you can determine the period by looking at the graph (Example 2). Notice in Figure 6.57 that you get the period by looking at one complete cycle, from high to low and back to high. Be sure to follow the computations in solving for *B*. It is easy to make a mistake when the variable is in the denominator.

- Notice in Example 3 that you are given the frequency (60 cycles per second), not the period. Be careful solving the equation for *B*. You may want to write down more steps than are shown.

- Recall also what you learned about horizontal shifts in Chapter 5 as you read about the parameter *h* (beginning in the middle of p. 271). Be sure to notice the warning about factoring to compute *h* on p. 271. You need to heed this warning in Example 5, p. 272.

- In Examples 7 and 8 you are asked to find a formula that models sinusoidal behavior. In Example 7, the behavior is given by a graph, but in Example 8 you have a verbal description. Notice that the graph in Figure 6.63 shows exactly one complete cycle, from the low to the high and back to the low again, so the period is 24.

- Study Example 8 carefully. It may help to sketch a graph first which fits the description of the population cycle. When you draw the graph, it must be concave up at the low point, changing to concave down at the midline. The graph should cross the midline a quarter of the way through the cycle (at $t = 3$) and again at $t = 9$. Notice that the midline is the average of the maximum and the minimum population, and that the amplitude is half their difference. Notice also that a negative cosine function is used, since the graph is at its low point when $t = 0$. Be careful with the units on the t-axis. We usually think of January as the first month, February as the second month, and so on, but the numbers we use in dates are not the numbers that correspond to the months in the graph. Since $t = 0$ corresponds to January 1, then $t = 1$ corresponds to February 1, and so on. So July 1 corresponds to $t = 6$, December 1 to $t = 11$, and back to January 1 again when $t = 12$.

- Recall Example 6, p. 255, where you expressed the height of the London ferris wheel as a function of the angle θ. In Example 9, p. 275, you get the height as a function of t, the number of minutes after boarding. As in Example 8, p. 274, it helps to sketch the graph first. Notice that the graph starts at the low point, since you boarded at the 6 o'clock position. Recall that in Examples 7 and 8, we used a negative cosine function when the graph started at the low point. You could use a negative cosine in this example as well. (See Problem 32, p. 277.) But we can also think of the graph as a sine graph that has been shifted horizontally, and we get a formula that is similar to the one in Example 6, p. 255.

- Understand the distinction between the terms **horizontal shift,** p.271, and **phase shift**, p. 273. From the sketch of a graph, the horizontal shift is more apparent. However, you may hear the term *phase shift* in applications, and you should understand what it means. Notice in Fig 6.61 that the horizontal shift of $\pi/12$ corresponds to a phase shift of $\pi/4$.

REVIEWING THE BASICS

You should be able to:

- Determine the amplitude, period, midline, and horizontal shift of any function of the form $y = A \sin B(t - h) + k$ or $y = A \cos B(t - h) + k$, and sketch the graph without using a calculator or computer.

- Write a formula for a given sinusoidal graph.

- Write a function that models sinusoidal behavior given in a verbal description.

Practice Problems

1. How is the graph of $y = \sin(x + \pi/2)$ related to the graph of $y = \sin x$? ______________

   ____ Sketch the graph of $\sin(x + \pi/2)$.

   (You should be able to do this without using a graphing calculator or computer).

   You should recognize the graph. It is the same as the graph of $y =$ __________.

2. State the amplitude, period, and midline for each function, and sketch the graph without using a calculator or computer. Label the t-coordinates where the graph changes direction and where the graph crosses the midline.

   (a) $y = 150\cos(60\pi t)$ (b) $y = 5 - 3\sin(2t)$

3. Find a possible formula for the graph.

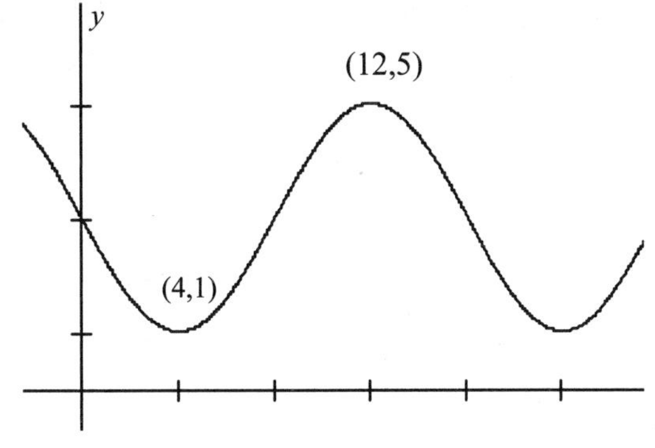

4. A population of animals oscillates sinusoidally from a low of 800 on January 1 to a high of 1500 on July 1 and then back to 800 again on January 1.

   (a) Sketch a graph of the population for one year, starting with $t = 0$ on January 1, and measuring t in months.

   (b) Find a possible formula for the population P at time t.

Solutions to Practice Problems

1. The graph of $y = \sin x$ is shifted left $\pi/2$ units. It is the same as the graph of $y = \cos x$.

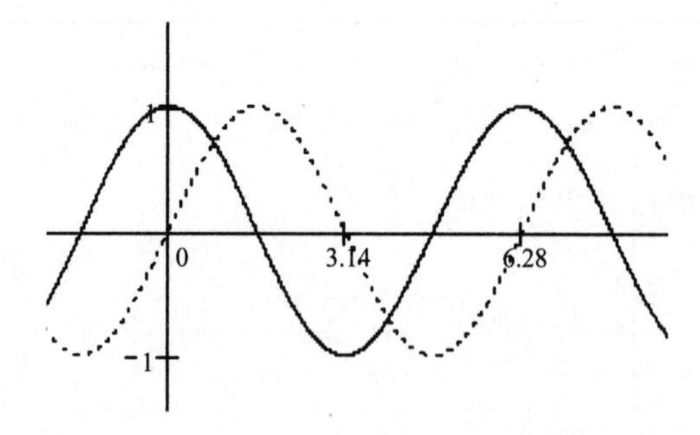

2. (a) Amplitude 150, period $\dfrac{2\pi}{60\pi} = \dfrac{1}{30} = 0.0333$; midline $y = 0$

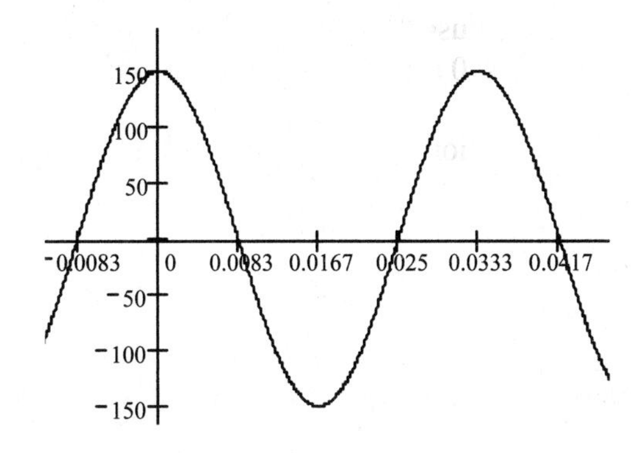

(b) Amplitude 3, period $\frac{2\pi}{2} = \pi$; midline $y = 5$

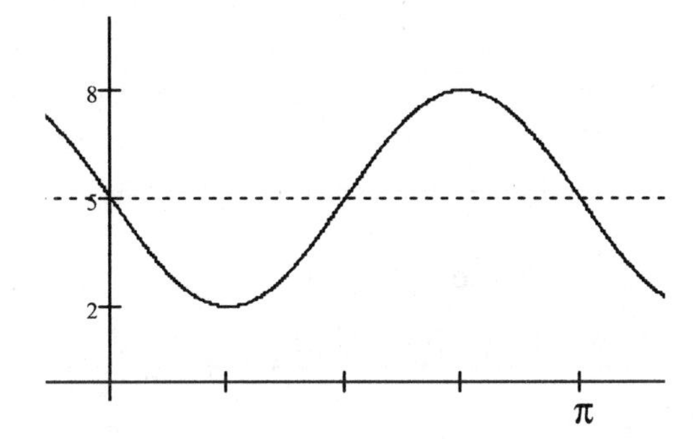

3. The amplitude is 2; the midline is $y = 3$; from $t = 4$ to $t = 12$ is half of a complete cycle, so the period is $16 = 2\pi/B$, and $B = \pi/8$. The graph starts at the midline when $t = 0$, and is a sine curve reflected across the midline. A possible formula is $y = -2 \sin\left(\frac{\pi}{8}t\right) + 3$.

4. (a)

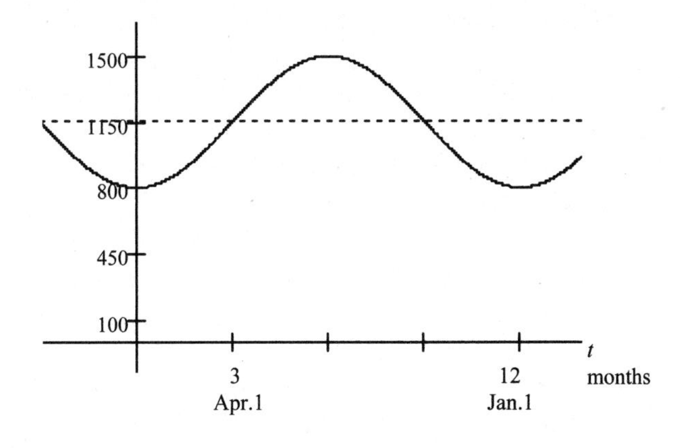

(b) midline: $P = (800 + 1500)/2 = 1150$; amplitude $= (1500 - 800)/2 = 350$; period $= 12 = 2\pi/B$, so $B = \pi/6$. The graph is a cosine graph that has been reflected about the midline. A possible formula is $P = -350\cos\left(\frac{\pi}{6}t\right) + 1150$.

MASTERING CONCEPTS AND SKILLS

Use the √, ?, * system on the exercises assigned by your instructor for this section.

Assigned Problems:

√

?

---

*

6.6 OTHER TRIGONOMETRIC FUNCTIONS

READING YOUR TEXTBOOK: Read Section 6.6, pp.279-283.

As you read:

- Learn the **definition of the tangent function** (p. 279) and form a mental picture of its graph (Figure 6.71). Notice that the tangent is not defined at $\pi/2$, $-\pi/2$, $3\pi/2$, $-3\pi/2$, and so on, and that the graph has vertical asymptotes at these values. Also note that the period is π.

- Learn the relationship **tan θ = sin θ/cos θ** (Box, p. 281). You can use this relationship to get the exact values of tan θ for the "special angles" and compare with the approximations given by your calculator. (See Example 2.)

- Learn the very important relationship $\cos^2\theta + \sin^2\theta = 1$, sometimes called the **Pythagorean identity** (Box, p. 281). Be sure to understand the justification of this relationship. Notice in Figure 6.69, p. 279, that when θ is acute, this relationship is the Pythagorean theorem that you know from geometry.

- Study Example 4, p.282, carefully to see a use of the Pythagorean identity. Notice especially why sin θ is negative.

- Learn the names and abbreviations of the reciprocals of the trigonometric functions. **Warning**: The keys $\sin^{-1}$, $\cos^{-1}$, and $\tan^{-1}$ on your calculator do not give these reciprocal functions. These keys give **inverse functions**, which we study in the next section. If you want to calculate sec θ, you have to type $1/\cos\theta$, and so on.

REVIEWING THE BASICS

You should be able to:

- State the definition of tangent of θ, and the domain and range of the tangent function.
- Interpret $\tan\theta$ as the slope of a line. (See Figure 6.69.)
- Sketch a graph of the tangent function without using a calculator, and label the intercepts and asymptotes.
- Find $\tan\theta$ if you know $\sin\theta$ and $\cos\theta$.
- Use the Pythagorean identity as in Example 4, p. 282.
- Use the graphs of $\sin\theta$, $\cos\theta$, and $\tan\theta$ to sketch graphs of $\csc\theta$, $\sec\theta$, and $\cot\theta$.

Practice Problems

1. Find the tangent of the angle θ shown in the sketch.

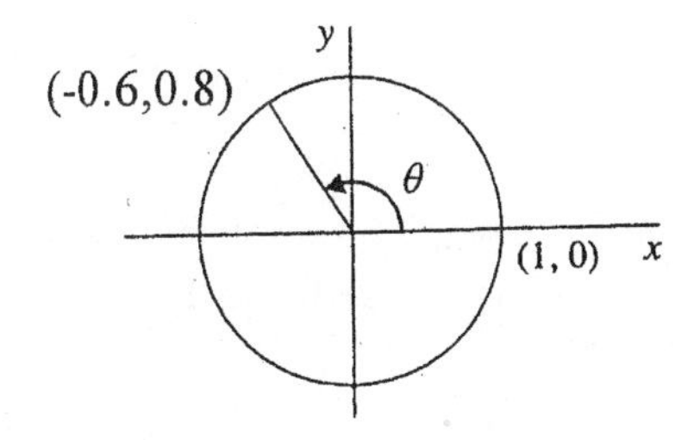

2. Suppose $\sin\theta = .4$ and $\pi/2 < \theta < \pi$. Use the Pythagorean identity to find $\cos\theta$.

3. Find $\tan\theta$ for the angle θ in Problem 2.

Answers to Practice Problems

1. $\tan\theta = 0.8/(-0.6) = -4/3$
2. $\sin^2\theta + \cos^2\theta = 1$; $0.4^2 + \cos^2\theta = 1$; $\cos^2\theta = 1 - 0.4^2$ and $\cos\theta < 0$ (since $\pi/2 < \theta < \pi$), so $\cos\theta = -\sqrt{1-.4^2} \approx -0.92$
3. $\tan\theta = \sin\theta/\cos\theta = 0.4/(-0.92) \approx -0.44$

MASTERING CONCEPTS AND SKILLS

Use the √ , ?, * system on the exercises assigned by your instructor for this section.

Assigned Problems:

√

?

---

*

6.7 INVERSE TRIGONOMETRIC FUNCTIONS

READING YOUR TEXTBOOK: Read Section 6.7, pp. 285-292.

As you read:

- Use your graphing calculator or computer as you work through Example 1. Set your calculator in radian mode. You can also solve the equation by graphing the line $y = .4$ and finding the points of intersection with the graph of $y = \cos t$.

- Learn what the **inverse cosine**, or $\mathbf{cos^{-1}}$ key on your calculator does. Remember: it does not give the reciprocal of the cosine, which is the secant.

- Use your graphing calculator or computer to graph the function in Example 2, p. 286. Set the viewing window as shown in Figure 6.79, and be sure your calculator is in radian mode. You can solve the equation by tracing or by finding the point of intersection with the line $y = 12{,}000$.

- Follow the computations in Example 2, where you solve the same problem using the inverse cosine. Be sure you are using your calculator correctly to evaluate $\cos^{-1}(-0.4)$. Notice how you use symmetry to get the second solution.

- **Memorize** the definition and the notation of the **inverse cosine function** (Box, p. 287). Notice especially the **domain [-1, 1]** and **range** $\mathbf{[0, \pi]}$.

- Notice that Example 3 is done without a calculator. You just need to keep in mind a mental picture of the cosine graph and the range of the arccosine. Compare the answers with the answers given by your calculator. Also notice that both notations $\cos^{-1}$ and arccos are used in this example.

- Notice the warning about the notation that follows Example 3, and the illustration of the difference between $\cos^{-1}y$ and $(\cos y)^{-1}$.

- **Memorize** the definition and notation of **inverse sine** and **inverse tangent** (Box, p. 289). Notice the **domain** and **range** of each function.

- Notice again that most of Examples 4 and 5 are done by knowing the graphs of sine and tangent, as well as the range of their inverse functions.

- Follow the algebra and the use of the arcsin function to solve the ferris wheel problem in Example 6, pp.289-290. Try it also by graphing $f(t)$ and the line $y = 400$, and determining when the graph of f is above the line.

- Learn what is meant by a **reference angle** (Box, p. 291). Remember that the circles in Figures 6.84-6.86 have radius 1, so the coordinates of P are $(\cos 65^\circ, \sin 65^\circ)$. You can use the coordinates of P to get the coordinates of the point associated with $180^\circ - 65^\circ$ in the second quadrant, $180^\circ + 65^\circ$ in the third quadrant, and $360^\circ - 65^\circ$ in the fourth quadrant. (See Example 7.)

- Notice how you use reference angles to get multiple solutions to trigonometric equations in Examples 8 and 9. Notice that in Example 8, we had $\cos \theta > 0$, so we looked for solutions in the first and fourth quadrants. But in Example 9, $\cos \theta < 0$, so we look for solutions in the second and third quadrants. Notice also that the answer was asked for in degrees, so your calculator needs to be in degree mode.

- Set your calculator in radian mode for Example 10. (The function giving the height is a transformation of the function $y = \sin t$, which has period 2π. See Example 9, p. 275). In general, if the reference angle for a trig equation is θ in radians, then second quadrant solutions are $\pi - \theta$, third quadrant solutions are $\pi + \theta$, and fourth quadrant solutions are $2\pi - \theta$.

REVIEWING THE BASICS

You should be able to:

- Solve trigonometric equations by graphing.

- State the definitions, including the range, of the **inverse sine, inverse cosine, and inverse tangent** functions. Also recognize and use both notations for each function.

- Solve trigonometric equations by using inverse trigonometric functions, and use reference angles to get all the solutions.

Practice Problems

1. (a) Solve by graphing: $\sin t = 0.8,\ 0 \leq t \leq 2\pi$. (Note: There are two solutions.)

   (b) Solve the same equation using the inverse sine.

2. Find the coordinates of the points Q, R, and S. The angles with the x-axis are all equal in magnitude.

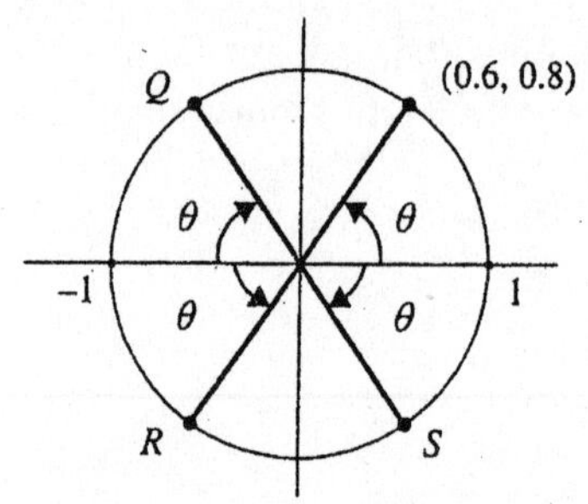

3. Explain why the question "Find arccos 2" does not make sense. ____________________

   ____

4. Refer to Example 7, p.274. During what interval of time is the temperature of the surface of the water above 19°C?

Solutions to Practice Problems

1. (a) Graph $y = \sin t$ and $y = 0.8$.

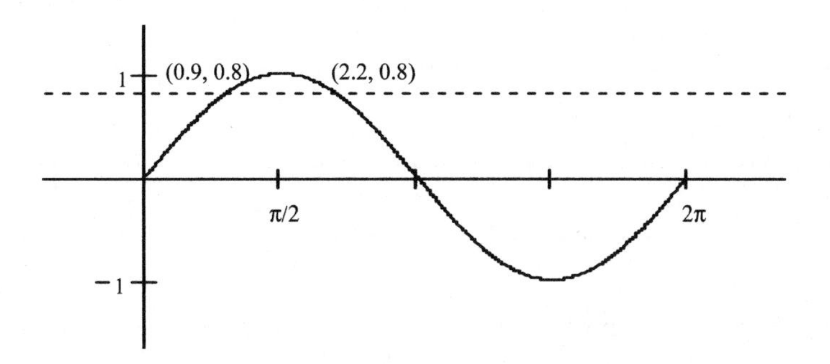

The calculator shows that the graphs intersect when $t = 0.927$ and when $t = 2.214$

(b) Since $0.8 > 0$, there are solutions in the first and second quadrants. The two solutions are $t = \sin^{-1} 0.8 = 0.927$ and $t = \pi - 0.927 = 2.214$

2. $Q = (-0.6, 0.8)$; $R = (-0.6, -0.8)$; $S = (0.6, -0.8)$
3. $\arccos 2 = \theta$ means that $\cos \theta = 2$, which is not possible, since $-1 \leq \cos \theta \leq 1$.
4. Solving by graphing: Graph $y = f(t)$ and $y = 19$. The calculator shows the two graphs intersect when $t \approx 4.7$ and when $t \approx 19.3$.

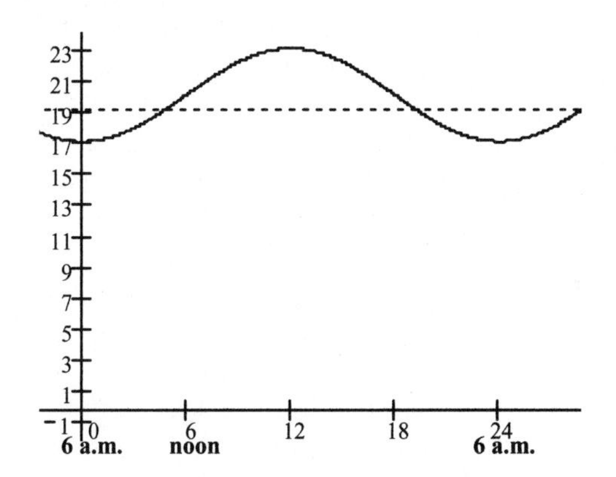

Using inverse cosine: First solve $-3\cos\left(\frac{\pi}{12}t\right) + 20 = 19$; $-3\cos\left(\frac{\pi}{12}t\right) = -1$;

$\cos\left(\frac{\pi}{12}t\right) = \frac{1}{3}$; $\frac{\pi}{12}t = \cos^{-1}\frac{1}{3} = 1.23$ or $\frac{\pi}{12}t = 2\pi - 1.23 = 5.05$

So $t = 1.23(12/\pi) \approx 4.7$, or $t = 5.05(12/\pi) \approx 19.3$

So $t = 4.7$ hrs after 6 am, or about 10:42 am, and $t = 19.3$ hrs after 6 am, or about 1:18 am the next day. Looking at the graph, we see that the temperature is above 19°C between 10:42 am and 1:18 am the next morning.

MASTERING CONCEPTS AND SKILLS

Use the √ , ?, * system on the exercises assigned by your instructor for this section.

Assigned Problems:

√

?

---

*

CHAPTER SEVEN

TRIGONOMETRY

7.1 GENERAL TRIANGLES: LAWS OF SINES AND COSINES

READING YOUR TEXTBOOK: Read Section 7.1, pp. 308-311.

As you read:

- Be careful with the notation. The small letters stand for the length of the sides, and the capital letters stand for the angle opposite the side with the same letter.
- Recall that the sum of the angles in a triangle is 180°; so if you know two of the angles, you can determine the third.
- Note that the Law of Cosines is a generalization of something you already know; the Pythagorean theorem is the Law of Cosines when c is the hypotenuse and $C = 90^{\circ}$. See the discussion on p. 308.
- Think about what you know from Euclidean geometry. For instance, knowing side-angle-side completely determines a triangle. The Law of Cosines gives you a method of actually finding the third side.
- The Law of Sines reminds you that triangles with the same angles are similar, and so knowing one side determines the other two sides. (You must know which angle is opposite the given side.)
- Note that Example 4 is showing that side-side-angle does not determine a unique triangle.
- Refer to the graph of $y = \sin\theta$ on p. 264 and the definition of $\sin^{-1} y$ on p. 289 to help you understand the computations in Example 4.

Practice Problems

1. What are the angles of a triangle whose sides are 5, 7, and 9? Which angle is opposite the side of length 7?

2. If the angles of a triangle are A, B, and C, what is C if $A = 73°$ and $B = 42°$?

3. The angles of a triangle are $A = 37°$, $B = 82°$, and $C = 61°$.

   (a) If the side opposite A is 8 cm, what are the other two sides?

   (b) If the side opposite B is 8 cm, what are the other two sides?

4. Use the triangle to answer the two questions that follow. (Figure not to scale)

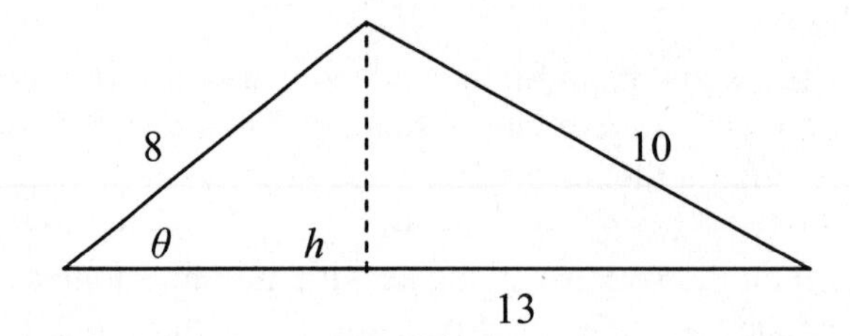

   (a) What is θ?

   (b) What is h?

Solutions to Practice Problems

1. Use $c^2 = a^2 + b^2 - 2ab \cos C$, with $a = 5$, $b = 7$, and $c = 9$. So $81 = 25 + 49 - 70 \cos C$

   $7 = -70 \cos C$; so $\cos C = -.1$; $C = \cos^{-1}(-.1) \approx 95.7^\circ$ or 1.67 radians. Now use $a = 5$, $b = 9$, and $c = 7$. Then $49 = 25 + 81 - 90 \cos C$, or $\cos C = 57/90 \approx .633$; $C = \cos^{-1}(.633) \approx 50.7^\circ$ or .88 radians. The third angle is $(180 - 95.7 - 50.7)^\circ = 33.6^\circ$, or $\pi - 1.67 - .88 = .59$ radians. (For practice, you may want to use the Law of Cosines again to compute the third angle.) When we used $c = 7$ we found that the opposite angle was 50.6° or .88 radians.
2. $C = (180 - 73 - 42)^\circ = 65^\circ$
3. (a) Use $\dfrac{\sin A}{a} = \dfrac{\sin B}{b}$ with $a = 8$. Then $\dfrac{\sin 37^0}{8} = \dfrac{\sin 82^0}{b}$, or $b = \dfrac{8 \sin 82^0}{\sin 37^0} \approx 13.2$ cm. To find c, use $\dfrac{\sin 37^0}{8} = \dfrac{\sin 61^0}{c}$, so $c \approx 11.6$ cm. The other two sides have length 13.2 cm and 11.6 cm. Notice that the longer sides are opposite the larger angles.

   (b) Now we use $\dfrac{\sin 37^0}{a} = \dfrac{\sin 82^0}{8}$, so $a \approx 4.9$. Finally, $\dfrac{\sin 61^0}{c} = \dfrac{\sin 82^0}{8}$, so $c \approx 7.1$. The other two sides have length 4.9 cm and 7.1 cm, with the longer sides opposite the larger angles.
4. (a) Use the Law of Cosines with $a = 8$, $b = 13$, and $c = 10$, with θ opposite c. Then $100 = 64 + 169 - 208 \cos \theta$, so $\cos \theta \approx .639$, and $\theta = \cos^{-1}(.639) \approx 50.3^\circ$ or .877 radians.

   (b) $\sin \theta = h/8$, so $h = 8 \sin 50.3^\circ \approx 6.2$

MASTERING CONCEPTS AND SKILLS

Use the √ , ?, * system on the exercises assigned by your instructor for this section.

Assigned Problems:

√

?

---

*

7.2 TRIGONOMETRIC IDENTITIES

READING YOUR TEXTBOOK: Read Section 7.2, pp. 313-318.

As you read:

- Understand what is meant by an **identity** as opposed to an equation. Examples 2 and 3 show why identities are useful.

- You should already know the **tangent identity** and the **Pythagorean identity** (Boxes, p. 314).

- Notice the use of the tangent identity and the inverse tangent function in Example 2. Be sure your calculator is in radian mode. Note, too, that $\tan\theta > 0$, so there is also a third quadrant solution. For practice, solve this equation by graphing $y = 2\sin x$ and $y = \sqrt{2}\cos x$ for $0 \le x \le 2\pi$ and finding the points of intersection.

- Be sure to see the important note in the solution to Example 2 about dividing both sides of an equation by a variable quantity. You risk losing solutions to the equation. For example, if you divide both sides of the equation $x^2 = x$ by the variable x, you get $x^2/x = x/x$, or $x = 1$ as the solution. But $x = 0$ is also a solution to the original equation.

- Notice in Example 3 how an alternate form of the Pythagorean identity transforms the given equation into a quadratic equation in $\cos t$. Work through all the computations.

- Notice the differences between $\sin 2\theta$ and $2\sin\theta$. Think about the amplitude and period of each one, and look at their graphs.

- Follow the derivation of the **formula for sin 2*θ***, pp. 315-316. It shows a clever use of the Law of Sines, and some right triangle trigonometry. This identity is worth memorizing (Box, p. 316).

- Note that the use of the double-angle formula in Example 4 transforms the original equation into one that can be factored. You should know the solutions to $\sin t = 0$ because you know the intercepts of the sine graph. If you have not memorized the trig functions at the "special angles," you can use the inverse cosine and reference angles to get the solutions to $\cos t = ½$. You can check your solutions by graphing $y = \sin 2x$ and $y = \sin x$ for $0 \le x \le 2\pi$ and finding points of intersection.

- Read the derivation of the double-angle formula for cosine on pp.316-317 to see an interesting use of the Pythagorean identity and the double-angle formula for sine. Note that the Pythagorean identity gives two other forms of the double-angle formula for cosine (Boxes, p.317.).

- Notice other identities that you can observe from looking at the graphs. The cosine graph is symmetric about the y-axis, so the cosine is an even function. The graphs of sine and tangent are symmetric about the origin, so these are odd functions. You can also get identities by observing that a cosine graph is a horizontal shift of a sine graph.

- Follow your instructor's directions about memorizing these formulas.

REVIEWING THE BASICS

You should be able to:

- Use the identities given in the boxes to rewrite expressions.

- Use the identities to verify other identities.
- Use the identities to solve equations.
- Use the identities to derive other formulas.
- Use the identities to evaluate expressions.
- State any identity your instructor has asked you to memorize.

Practice Problems

1. Describe the differences between $\sin 2\theta$ and $2 \sin \theta$. Specifically, give the amplitude and period of each function.

2. Use the tangent identity to find all solutions in $[0, 2\pi]$ to the equation $2 \sin t = 5 \cos t$. Check your solutions by graphing.

3. Find all solutions in $[0, 2\pi]$ to the equation $2 \cos^2 t = \sin t - 1$. Use a form of the Pythagorean identity to transform the equation into a quadratic equation in $\sin t$, and solve by factoring. Check your solution by graphing. Remember that to graph $y = 2 \cos^2 t$, you need to type $y = 2 (\cos t)^2$.

4. Suppose $\sin t = 0.85$ and $\pi/2 < t < \pi$.

   (a) Use the Pythagorean identity to find $\cos t$.

(b) Use the tangent identity to find tan t.

(c) Use the double-angle identities to find sin $2t$ and cos $2t$.

Solutions to Practice Problems

1. $\sin 2\theta$ has amplitude 1 and period π, while $2 \sin \theta$ has amplitude 2 and period 2π.
2. Assuming $\cos t \neq 0$, we can divide both sides by $2 \cos t$ to get $\dfrac{\sin t}{\cos t} = 2.5$, so $\tan t = 2.5$, and $t = \tan^{-1}(2.5) \approx 1.19$. Since $\tan t > 0$, there is also a third quadrant solution, $t = \pi + \tan^{-1}(2.5) \approx 4.33$. Note: When $\cos t = 0$, then the right-hand side of the equation equals 0, but the left-hand side of the equation equals 2 or -2, so the equation is not satisfied. Graphing $y = 2 \sin t$ and $y = 5 \cos t$ and finding points of intersection gives the same solutions.

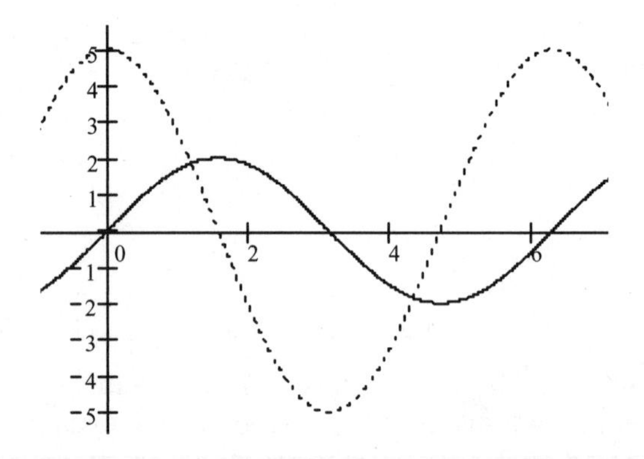

3. Use $\cos^2 t = 1 - \sin^2 t$. This gives $2(1 - \sin^2 t) - \sin t + 1 = 0$, or $-2 \sin^2 t - \sin t + 3 = 0$. Multiplying both sides by -1 gives $2 \sin^2 t + \sin t - 3 = 0$. Factoring the left-hand side gives $(2 \sin t + 3)(\sin t - 1) = 0$. So either $\sin t = -3/2$, which is not possible, or $\sin t = 1$. So $t = \pi/2$ is the only solution. Graphing $y = 2 (\cos t)^2$ and $y = \sin t - 1$ confirms the result.

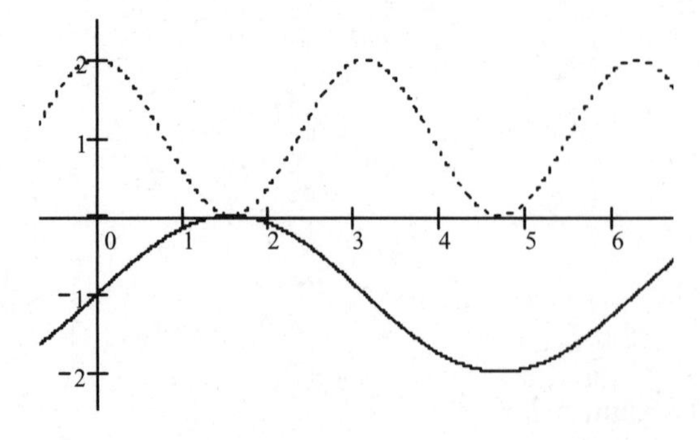

4. (a) $\cos^2 t = 1 - \sin^2 t = 1 - 0.85^2$ and $\cos t < 0$, so $\cos t = -\sqrt{1-.85^2} \approx -0.527$
   (b) $\tan t = \sin t / \cos t \approx 0.85/(-0.527) \approx -1.613$
   (c) $\sin 2t = 2 \sin t \cos t \approx 2(0.85)(-0.527) = -0.896$; $\cos 2t = 1 - 2\sin^2 t = 1 - 2(0.85)^2 \approx -0.445$

MASTERING CONCEPTS AND SKILLS

Use the √ , ?, * system on the exercises assigned by your instructor for this section.

Assigned Problems:

√

?

---

*

7.3 SUM AND DIFFERENCE FORMULAS FOR SINE AND COSINE

READING YOUR TEXTBOOK: Read Section 7.3, pp. 320-325.

As you read:

- Be aware of the need for these formulas—for example, that $\sin(\theta + \varphi) \neq \sin\theta + \sin\varphi$. (Use your calculator to see that $\sin 2 + \sin 3 = 1.05$, but $\sin(2 + 3) = \sin 5 = -0.959$.)

- Use your calculator to see that the sum formula works for some specific values. For example, writing $5 = 2 + 3$, and using the sum formula, we get $\sin(2 + 3) = \sin 2 \cos 3 + \cos 2 \sin 3 = (0.909)(-0.990) + (-0.416)(0.141) = -0.959 = \sin 5$.

- Notice that the sum formula (Box, p. 320) gives the double angle formula when $\theta = \varphi$.

- Follow the use of the sum formula in Example 1, p. 321. Recall that the values of sine and cosine for 30° and 45° can be found exactly from the special triangles. So the sum formula gets the exact value of sin 75°. Compare the calculator approximation of sin 75° to the exact value obtained in Example 1.

- Graph $y = \sin t + \cos t$ using your graphing calculator or computer in radian mode. Set the window approximately as in Figure 7.23. Notice that the graph does look like a sine wave with period 2π. Compare the t-intercept just to the left of the origin with the calculator approximation of $-\pi/4$. This suggests a horizontal shift of $\pi/4$ units to the left. Follow the use of the sum formula for $\sin(t + \pi/4)$, which results in a formula for sin t + cos t that only involves the sine function. Now graph the function $f(t)$ in the same viewing window. The graphs should coincide exactly.

- Graph the function $y = 2 \sin 3t + 5 \cos 3t$ in the viewing window [-3, 3] by [-7, 7]. Notice that it also looks like a sine wave. The discussion on p. 322 shows how to rewrite a function of this form as a sine function (Box, p. 322). Notice the use of the sum formula and the use of the Pythagorean identity. It is important that both functions in the sum have the same period, and hence the same parameter B. Follow the use of the formula to get the function $g(t)$ in Example 2. Now graph this new function in the same viewing window. The graphs should coincide (except for a small difference that results from using an approximation of arctan (5/2). You will see this difference if you trace on each graph).

- Follow your instructor's directions about which, if any, of the formulas derived in this section you should memorize.

- Notice how the identities are derived (pp. 323-325). Only one of them, cos $(\theta - \varphi)$, is derived using angles. Follow the use of the Law of Cosine to get the distance AB (Figure 7.24). Also remember the definition of the sine and cosine, which gives you the coordinates of the points A and B. Using the distance formula gives another way to write the distance AB. Notice two uses of the Pythagorean identity in simplifying AB^2, p. 324.

- The other identities are derived from this one. Notice clever substitutions, as well as the use of odd and even properties of sine and cosine.

REVIEWING THE BASICS

You should be able to:

- Use the formulas derived in this section to derive other identities.

- Transform the sum of a sine function and a cosine function with the same period into a single sine function. (Use box, p.322.)

Practice Problems

1. Using your calculator, illustrate with specific values that $\cos(u + v) \neq \cos u + \cos v$.

2. Use the same values you used in Problem 1 to illustrate that the sum-of-angle formula for cosine does give the correct value for $\cos(u + v)$.

3. Express $f(t) = 3 \sin 2t - 4 \cos 2t$ as a single sine function. Check by graphing both functions in the same viewing window and see if the graphs coincide. Sketch the graph. Indicate the period and the amplitude.

4. Verify that $\sin(3t) = \sin t\,(2 \cos t + 1)(2 \cos t - 1)$. (Hint: Write $3t = 2t + t$, and use the sum-of-angle formula for sine.)

Solutions to Practice Problems

1. One example would be $u = 4$, $v = 3$. Then $\cos(4 + 3) = \cos 7 = 0.7539$, but $\cos 4 + \cos 3 = -0.654 + (-0.990) = -1.644$.
2. Using the sum-of-angle formula for cosine gives $\cos(4 + 3) = \cos 4 \cos 3 - \sin 4 \sin 3 = (-0.654)(-0.990) - (-0.757)(0.141) = 0.754 = \cos 7$.
3. The two functions in the sum have the same period $2\pi/2 = \pi$, so use the formula in the Box, p. 322, with $a_1 = 3$ and $a_2 = -4$. So $A = \sqrt{9+16} = 5$, and $\tan \varphi = -4/3$. Since $\sin \varphi = -4/5$ and $\cos \varphi = 3/5$, it follows that φ is in the fourth quadrant, so $\varphi = \tan^{-1}(-4/3) = -0.927$. So $f(t) = 5 \sin(2t - 0.927)$. The graph has amplitude 5 and period π.

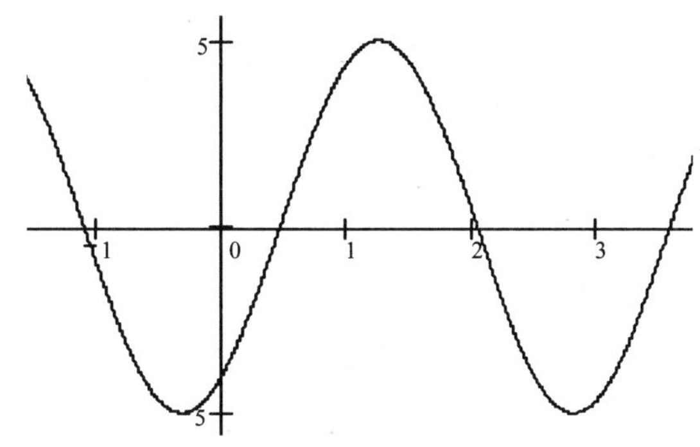

4. $\sin(3t) = \sin(2t + t) = \sin 2t \cos t + \cos 2t \sin t = (2 \sin t \cos t) \cos t + (2 \cos^2 t - 1)\sin t$
$= \sin t\,(2 \cos^2 t + 2 \cos^2 t - 1) = \sin t\,(4 \cos^2 t - 1) = \sin t\,(2 \cos t + 1)(2 \cos t - 1)$

MASTERING CONCEPTS AND SKILLS

Use the √ , ?, * system on the exercises assigned by your instructor for this section.

Assigned Problems:

√

?

---

*

7.4 TRIGONOMETRIC MODELS

READING YOUR TEXTBOOK: Read Section 7.4, pp. 327-339.

As you read:

- Be aware that Example 1 is an application of one of the formulas developed in the previous section. The formulas for the functions in this example let us add them using the formula in the Box p. 322.

- Contrast the situation in Example 1 with the two functions in Example 2, whose sum is not sinusoidal.

- Follow the discussion on “Acoustic Beats” to see an important application of adding functions with different periods.

- Remember that the trig functions are important in modeling periodic behavior. Several discussions in this section show that the trig functions can be important when combined with other functions to model behavior that has some periodic component.

REVIEWING THE BASICS

You should be able to:

- Model simple periodic behavior with a sinusoidal function.

- Describe in words a function that is not periodic but that has some periodic component.

Practice Problems

1. The population of field mice in a certain district is at its minimum value of 3000 on March 1. The population rises to its maximum of 7000 on September 1 and then decreases over the next six months to 3000 again. Find a sinusoidal function that could be a possible formula for $P(t)$, the population of field mice, as a function of t, the time in months since January 1.

2. Let $f(x) = \dfrac{\sin x}{x}$, $x \geq \pi/2$

   (a) Is f periodic?

   (b) Are the zeros of f periodic? Can you find a formula for the zeros of f?

   (c) Describe the behavior of $f(x)$ as $x \to \infty$.

Solutions to Practice Problems

1. We will use the form of the sinusoidal function as given on p. 261: $P(t) = A \sin (B(t - h)) + k$. The average of 3000 and 7000 is 5000, so $k = 5000$. The maximum value 7000 is 2000 greater than 5000, so the amplitude $A = 2000$. The value of B determines the period, which is 12 months. So $B = 2\pi/12 = \pi/6$. Finally, we want the maximum to occur when $t = 8$, so we can set $\sin((\pi/6)(8 - h)) = 1$, or $(\pi/6)(8 - h) = \pi/2$. Thus $h = 5$. Our answer is

$$P(t) = 2000 \sin (\frac{\pi}{6}(t-5)) + 5000 .$$

2. (a) No, f is not periodic.
   (b) Yes, the zeros are periodic with period π. The zeros of f are π, 2π, 3π, and so on, and so are given by the formula $n\pi$, where n is a positive integer.
   (c) The values of f oscillate between the graphs of $y = 1/x$ and $y = -1/x$. This function is another example of damped oscillation, which is discussed further in the text.

MASTERING CONCEPTS AND SKILLS

Use the √ , ?, * system on the exercises assigned by your instructor for this section.

Assigned Problems:

√

?

---

*

7.5 POLAR COORDINATES

READING YOUR TEXTBOOK: Read Section 7.5, pp.336-340.

As you read:

- Understand that polar coordinates and Cartesian coordinates are two different ways to specify the same point *P* in the plane.
- Understand what r and θ stand for; see the first paragraph of this section (p. 336).
- Learn how to change from polar coordinates to Cartesian coordinates; see the first bullet on p. 336.
- Learn how to change from Cartesian coordinates to polar coordinates; see the second bullet and the sentence following this bullet on p. 336.
- Remember the range of the inverse tangent (Box, p. 289).
- Discuss with your instructor if you can use technology to help you draw the graphs of polar equations.

REVIEWING THE BASICS

You should be able to:

- Find the Cartesian coordinates of a point *P* when given the polar coordinates.
- Find the polar coordinates of a point *P* when given the Cartesian coordinates.
- Identify the polar curve $r = k$ as a circle of radius k centered at the origin.
- Draw a simple graph of the form $r = f(\theta)$ for some function f.

Practice Problems

1. Find the Cartesian coordinates of P if the polar coordinates of P are given by $r = 5$ and $\theta = 2$.

2. Find polar coordinates for P if in rectangular coordinates $P = (-3, 2)$.

3. Describe in words the set of points whose polar equation is given by $r = 6$.

4. Graph the equation $r = 2 + \cos\theta$, for $0 \leq \theta \leq 2\pi$

Solutions to Practice Problems

1. Since $x = r\cos\theta$, we have $x = 5\cos 2 \approx -2.081$. (Remember if degrees are not explicitly written, then θ is measured in radians). $y = 5\sin 2 \approx 4.546$.

2. $r = \sqrt{(-3)^2 + 2^2} = \sqrt{9+4} = \sqrt{13}$. Note that P is in the second quadrant, so we want

   $\pi/2 < \theta < \pi$. These values of θ are contained in the range of the inverse cosine;

   $\cos^{-1}\left(\dfrac{-3}{\sqrt{13}}\right) = 2.553$ (Check that $\sin 2.553 = 0.555$ and that $\dfrac{2}{\sqrt{13}} = 0.555$ also). Thus

θ = 2.553. Because polar coordinates are not unique, we could also have $\theta = 2.553 + 2\pi$, or any multiple of 2π. (See Example 2, p.337.)

3. The graph of $r = 6$ is a circle with radius 6 centered at the origin.

4.

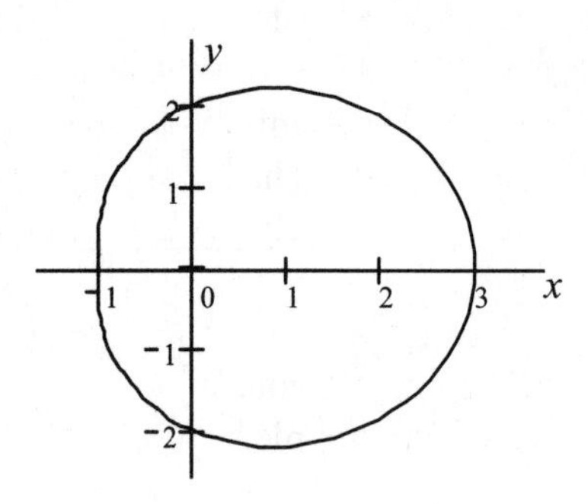

MASTERING CONCEPTS AND SKILLS

Use the √ , ?, * system on the exercises assigned by your instructor for this section.

Assigned Problems:

√

?

---

*

7.6 COMPLEX NUMBERS AND POLAR COORDINATES

READING YOUR TEXTBOOK: Read Section 7.6, pp. 341-347.

As you read:

- Don't be fooled by the word *imaginary*. Read the paragraph after the box on p.342; these numbers are used by engineers.

- Notice that the algebra of complex numbers is much like the algebra of polynomials, something that you are familiar with.

- You may want to quickly review Section 7.5, Polar Coordinates, to help you with Examples 4 and 5.

- Be aware of the paragraph of text between Examples 6 and 7. This paragraph can help you understand that these examples are just showing how one can use Euler's formula.

- Note that $\cos\theta + i\sin\theta$ is a point on the unit circle, and so we are led to the polar form of a complex number in the box on p. 345. The point with polar coordinates (r, θ) becomes the point $re^{i\theta}$. Example 11 shows how nicely some trig identities follow from this way of writing the point.

- Consult with your instructor about using your graphing calculator or other technology to perform arithmetic calculations with complex numbers.

REVIEWING THE BASICS

You should be able to:

- Find the conjugate of a complex number.

- Perform arithmetic calculations with complex numbers.

- Convert back and forth between the forms $z = x + iy$ and $z = re^{i\theta}$.

- Use the form $z = re^{i\theta}$ to find powers of z.

Practice Problems

1. (a) If $z = 3 - 2i$, what is the complex conjugate of z? ________________

   (b) If $z = 4e^{3i}$, what is the complex conjugate of z? ________________

2. Perform the indicated calculations and write the answer in the form $x + iy$.

   (a) $(7 - 4i) - (3 + 2i)$

   (b) $(6 + i)(5 - 4i)$

   (c) $\dfrac{2 - i}{4 + 3i}$

3. (a) Write $-2 + 2i$ in the form $re^{i\theta}$.

   (b) Write $5e^{2i}$ in the form $x + iy$.

4. Find a value for $\sqrt{5+12i}$.

Solutions to Practice Problems

1. (a) $3 + 2i$; (b) $4e^{-3i}$ (See explanation on p. 345 between the box and Example 8.)
2. (a) $(7 - 4i) - (3 + 2i) = 7 - 4i - 3 - 2i = 4 - 6i$; (b) $(6 + i)(5 - 4i) = 30 - 24i + 5i - 4i^2 =$ $30 - 19i - 4(-1) = 34 - 19i$; (c) The conjugate of $4 + 3i$ is $4 - 3i$, so $\dfrac{2-i}{4+3i} =$

   $$\frac{(2-i)(4-3i)}{(4+3i)(4-3i)} = \frac{5-10i}{25} = \frac{5}{25} - \frac{10}{25}i = 0.2 - 0.4i\,.$$

3. (a) $r^2 = (-2)^2 + 2^2 = 8$, so $r = \sqrt{8}$. The point is in the second quadrant and $\tan\theta = y/x$ $= -1$. Thus θ can be $3\pi/4$ or $3\pi/4 + 2n\pi$ for some integer n. Answer: $-2 + 2i = \sqrt{8}\, e^{i\,(3\pi/4)}$.

   (b) $5(\cos 2 + i \sin 2) = -2.081 + 4.56i$.
4. We first write $5 + 12i$ as $re^{i\theta}$; $r^2 = 5^2 + 12^2 = 169$, so $r = 13$. Next, $5 + 12i$ is in the first quadrant with $\tan\theta = 12/5$, so $\theta = \arctan(12/5) = 1.176$. So

   $$\sqrt{5+12i} = (5+12i)^{1/2} = (13e^{i(1.176)})^{1/2} = \sqrt{13}e^{i(0.588)} = \sqrt{13}\cos 0.588 + \sqrt{13}\sin 0.588 = 3+2i$$

   To check our answer we can compute $(3 + 2i)(3 + 2i) = 9 + 12i + 4i^2 = 9 + 12i - 4 = 5 +$ $12i$. Note that $1.176 + 2\pi$ is also in the first quadrant with the same terminal side, and so can be used for θ. What number do you get using $\theta = 1.176 + 2\pi$? (Ans: $-3 - 2i$; square this number to check it is a square root of $5 + 12i$.

MASTERING CONCEPTS AND SKILLS

Use the √ , ?, * system on the exercises assigned by your instructor for this section.

Assigned Problems:

√

?

---

*

CHAPTER EIGHT

COMPOSITIONS, INVERSES, AND COMBINATIONS OF FUNCTIONS

8.1 COMPOSITION OF FUNCTIONS

READING YOUR TEXTBOOK: Read Section 8.1, pp. 354-359.

As you read:

- Study carefully the discussion on p. 354 and Example 1 which follows. Learn what is meant by **composition of functions** (Box, p. 355).
- Follow the use of function notation and all the computations in Examples 2 and 3. Work through and understand each step. Read carefully; $f(g(x)) \neq f(x) \cdot g(x)$
- Refer to the graphs in Figure 8.1 as you follow each step in Example 4.
- Example 5 is extremely important—recognizing a complicated function as a composition of simpler functions. Notice that while other decompositions are possible, the most useful one has a polynomial as the "inside" function, the $g(x)$ function. Watch for functions of the form $e^{g(x)}$.

REVIEWING THE BASICS

You should be able to:

- Evaluate $f(g(x))$ from a table of values for f and g.
- Find a formula for $f(g(x))$ from the formulas for f and g, and use algebra to simplify the result.
- Express a function as the composition of simpler functions.

Practice Problems

1. The table below gives values of the functions f and g. Evaluate each expression. If there is not enough information given, state what additional information you would need to answer the question.

x	0	1	2	3	4	5
$f(x)$	3	5	0	2	1	4
$g(x)$	2	7	1	5	3	0

(a) $h(3)$ if $h(x) = g(f(x))$

(b) $k(5)$ if $k(x) = f(g(x))$

(c) $k(1)$ if $k(x) = f(g(x))$

2. Let $f(x) = \dfrac{1}{x+2}$ and $g(x) = x - 5$. Let $h(x) = f(g(x))$ and $k(x) = g(f(x))$.

(a) Evaluate $h(0)$.

(b) Evaluate $k(-3)$.

(c) What happens when you try to evaluate $h(3)$?

(d) Find a formula for $h(x)$ and simplify it.

(e) Find a formula for $k(x)$ and simplify it.

3. Let $R(x) = \sqrt{2+x^2}$. Find functions u and v so that $R(x) = u(v(x))$.

Solutions to Practice Problems

1. (a) $h(3) = g(f(3)) = g(2) = 1$
   (b) $k(5) = f(g(5)) = f(0) = 3$
   (c) $k(1) = f(g(1)) = f(7)$, which is not given in the table.
2. (a) $h(0) = f(g(0)) = f(-5) = \dfrac{1}{-5+2} = -\dfrac{1}{3}$

   (b) $k(-3) = g(f(-3)) = g(\dfrac{1}{-3+2}) = g(-1) = -1-5 = -6$

   (c) $h(3) = f(g(3)) = f(-2) = \dfrac{1}{0}$, which is not defined. So 3 is not in the domain of h.

   (d) $h(x) = f(g(x)) = f(x - 5) = \dfrac{1}{x-5+2} = \dfrac{1}{x-3}$

   (e) $k(x) = g(f(x)) = g\left(\dfrac{1}{x+2}\right) = \dfrac{1}{x+2} - 5 = \dfrac{1-5(x+2)}{x+2} = \dfrac{-5x-9}{x+2}$
3. Let $v(x) = 2 + x^2$ and $u(x) = \sqrt{x}$. Then $u(v(x)) = u(2 + x^2) = \sqrt{2+x^2} = R(x)$.

MASTERING CONCEPTS AND SKILLS

Use the √ , ?, * system on the exercises assigned by your instructor for this section.

Assigned Problems:

√

?

---

*

8.2 INVERSE FUNCTIONS

READING YOUR TEXTBOOK: Read Section 8.2, pp. 362-370.

As you read:

- Review the definition of log x (p. 152) and ln x (p. 155), and of the inverse trig functions (pp. 287-289).

- Learn how the **inverse function** is defined (Box, p. 362). Memorize the statement *exactly*. Then study Example 2, p. 362, for some practice using the definition.

- Notice the interpretations in Example 3. Also notice that you have to read the graph "backwards" to find $f^{-1}(25)$.

- Remember: In function notation, $f^{-1}(x)$ *does not mean* $1/f(x)$.

- Notice in Example 3 part (b) that you used algebra to find $f^{-1}(25)$. Now, in the middle of p. 363, you do the same calculations for a general P, and you get a formula for $t = f^{-1}(P)$.

- Study Example 4 carefully. You will need to remember how to use logs to solve exponential equations.

- WARNING: The algebra of finding inverse functions can be very challenging (See exercises 15-27, p.370). You will get better at these with careful practice.

- See in Example 6 how you can use the graph of f to find values of f^{-1}, even when you can't find a formula.

- Be aware that not all functions have inverses. Learn the **Horizontal Line Test** (Box, p. 365).

- Recall from Section 4.3 that the graph of $y = \ln x$ is the reflection of the graph of $y = e^x$ across the line $y = x$ (p. 170). Example 7, p. 366, shows this same property for the graph of $y = 2^x$ and its inverse. The discussion on pp. 367-368 generalizes this property to any invertible function. The box on p. 367 summarizes the relationships between the domain and range of a function and its inverse.

- Recall that $\ln e^x = x$ for all real numbers x and that $e^{\ln x} = x$ for $x > 0$. This "inverse" property is also true for any invertible function and its inverse (box, p. 368). Use this property in Example 8, p. 368, to decide whether two functions are inverses. Be sure to follow all the computations in this example.

REVIEWING THE BASICS

You should be able to:

- Interpret $f^{-1}(y)$ when $y = f(x)$ models physical quantities.
- Decide by looking at its graph if a function is invertible.
- Find values of f^{-1} from a graph or table of values for f.
- Find a formula for the inverse of a given function.
- Use the inverse properties (box, p.368) to decide if two functions are inverses.
- Refresh your memory about the domain and range of the three inverse trig functions discussed on p. 287-289.

Practice Problems

1. Suppose $C = f(x)$ is the cost (in dollars) of producing x items. Interpret each statement:

   (a) $f(500) = 1100$ ____

   (b) $f^{-1}(1775) = 950$ ____

   (c) If $C = f(x) = 350 + 1.5x$, find a formula for $x = f^{-1}(C)$.

2. Explain specifically why the function whose graph is shown below is not invertible.

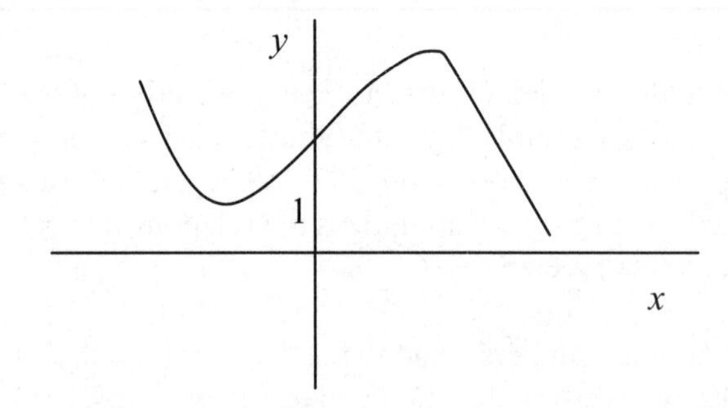

3. Use the inverse property to show that the functions $f(x) = 2x - 5$ and $g(x) = .5x + 2.5$ are inverses.

4. Suppose that f is invertible, and that $f(3) = 4$ and $f^{-1}(5) = 2$. Evaluate each expression. If the given information is insufficient, write unknown.

   (a) $f(2)$ (b) $f^{-1}(4)$ (c) $(f(3))^{-1}$ (d) $f^{-1}(f(3))$

Solutions to Practice Problems

1. (a) It costs $1100 to produce 500 items. (b) For $1775, you can produce 950 items
   (c) $C = 350 + 1.5x$ so $x = (C - 350)/1.5 = f^{-1}(C)$.
2. The graph fails the horizontal line test. For example, the line $y = 2$ meets the graph more than once, so $f^{-1}(2)$ is not well-defined.
3. $f(g(x)) = f(.5x + 2.5) = 2(.5x + 2.5) - 5 = x + 5 - 5 = x$
   $g(f(x)) = g(2x - 5) = .5(2x - 5) + 2.5 = x - 2.5 + 2.5 = x$
4. (a) $f(2) = 5$; (b) $f^{-1}(4) = 3$; (c) $(f(3))^{-1} = 1/f(3) = ¼$ (d) $f^{-1}(f(3)) = 3$

MASTERING CONCEPTS AND SKILLS

Use the √ , ?, * system on the exercises assigned by your instructor for this section.

Assigned Problems:

√

?

*

8.3 COMBINATIONS OF FUNCTIONS

READING YOUR TEXTBOOK: Read Section 8.3, pp.373-378.

As you read:

- Think about why it sometime makes sense to combine functions algebraically in physical situations. The first example deals with food supply and population (pp. 373-375). There are no difficult computations in this example. Just read for an understanding of why it makes sense to subtract these functions and how the graph of their difference is related to the graphs of the functions (Figures 8.20 and 8.21). On pp. 376-377, see why it makes sense to divide these functions. The paragraph about when shortages occur with each model is especially interesting (p.377). The discussion of crime rates (pp. 377-378) shows another situation where dividing functions produces significant information.

- Notice again the emphasis on vertical distances in Example 1, pp. 375-376.

- In Example 2, be aware of an important point in finding zeros of a function by factoring: an exponential expression a^x is never equal to zero. For the product to equal zero, one of the other factors must equal zero.

REVIEWING THE BASICS

You should be able to:

- Make a table of values for algebraic combinations of functions from a table of values for the functions.

- Find a formula for algebraic combinations of functions from formulas for the functions.

- Given graphs of two functions f and g, sketch a graph of $f + g$ or $f - g$.

Practice Problems

1. Complete the table.

x	0	1	2	3	4
$f(x)$	-3	5	4	-1	9
$g(x)$	2	8	-5	4	3
$f(x) + g(x)$					
$f(x)/g(x)$					

2. $r(x) = 1 + x^2$ and $s(x) = x - 2$.

   (a) Find $t(3)$ if $t(x) = r(x) - s(x)$

   (b) Find $q(-1)$ if $q(x) = r(x)s(x)$

   (c) Find formulas for $t(x)$ and $q(x)$.

Solutions to Practice Problems

1.

x	0	1	2	3	4
$f(x)$	-3	5	4	-1	9
$g(x)$	2	8	-5	4	3
$f(x) + g(x)$	-1	13	-1	3	12
$f(x)/g(x)$	-3/2	5/8	-4/5	-1/4	9/3=3

2. (a) $t(3) = r(3) - s(3) = 1 + 3^2 - (3 - 2) = 10 - 1 = 9$
   (b) $q(-1) = r(-1)s(-1) = 2\cdot(-3) = -6$
   (c) $t(x) = r(x) - s(x) = 1 + x^2 - (x - 2) = x^2 - x + 3$
   $q(x) = r(x)s(x) = (1 + x^2)(x - 2) = x + x^3 - 2 - 2x^2 = x^3 - 2x^2 + x - 2$

MASTERING CONCEPTS AND SKILLS

Use the √ , ?, * system on the exercises assigned by your instructor for this section.

Assigned Problems:

√

?

---

*

CHAPTER NINE

POLYNOMIAL AND RATIONAL FUNCTIONS

9.1 POWER FUNCTIONS

READING YOUR TEXTBOOK: Read Section 9.1, pp. 388-392.

As you read:

- Learn the meaning of the phrases "is directly proportional to" and "is inversely proportional to". See Examples 1 and 2 p. 388, and Example 4 p.392, as well as the boxes on p.388.

- Learn to recognize a power function. See box and Example 3, p. 389.

- Notice the change from radical notation to exponent notation in Example 3, parts (a) and (c). Review the definitions and properties of exponents, p.146, if you have forgotten them.

- Form mental pictures of the graphs of $y = x^p$ for p positive and even, positive and odd, negative and odd, or negative and even. (See Figures 9.4-9.7.)

- Notice that if $p < 0$, then x^p is very small when x is large. (See Table 9.2, p.391) Thus the x-axis is a horizontal asymptote. Again, if $p < 0$, then x^p is not defined when $x = 0$, and the y-axis is a vertical asymptote. (See Table 9.3, p. 391.)

- Remember that $\sqrt{x}$ is defined only for $x \geq 0$. The same is true for $y = x^p$ where $p = 1/n$ and n is even. The graphs of these functions are in the first quadrant. However, negative numbers do have cube roots. Similarly, functions of the form $y = x^p$ where $p = 1/n$ and n is odd are defined for all real numbers. Their graphs are in the first and third quadrants. (See Figures 9.8 and 9.9.) The graphs are concave down in the first quadrant.

- Notice that every function of the form $y = x^p$ passes through the point (1,1). When p is positive, the graph also contains the origin. If $y = kx^p$, then the point $(1, k)$ is on the graph.

- Study Example 5 carefully to see how to find the formula for a power function if you know two data points. Work through the computations.

REVIEWING THE BASICS

You should be able to:

- Match power functions with graphs without using a calculator.
- Describe the behavior of power functions for large x and for x near zero.
- Use algebra to find a formula for a power function if you know two data points.
- Translate "is proportional to" into a mathematical equation.

Practice Problems

1. Use the following graph to answer the questions. Assume $y = kx^p$, where p is an integer. Give a brief reason for your answer.

   (a) Is p odd or even?
   (b) Is p positive or negative?
   (c) What is the value of k?

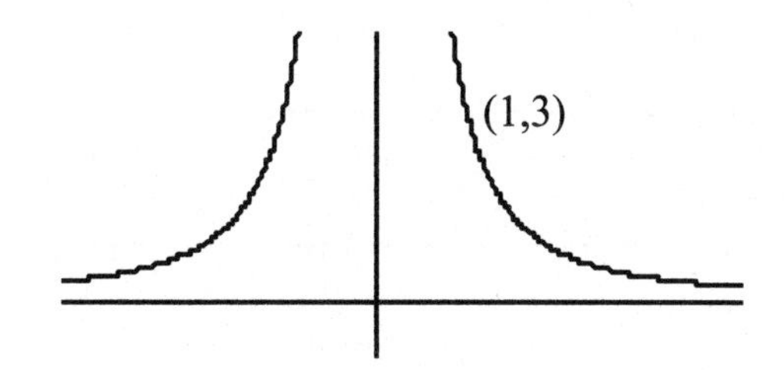

2. Without a calculator, match the following functions with the graphs. Explain briefly how you arrived at your answer.

   (i) $y = x^2$ (ii) $y = x^4$ (iii) $y = x^{1/2}$

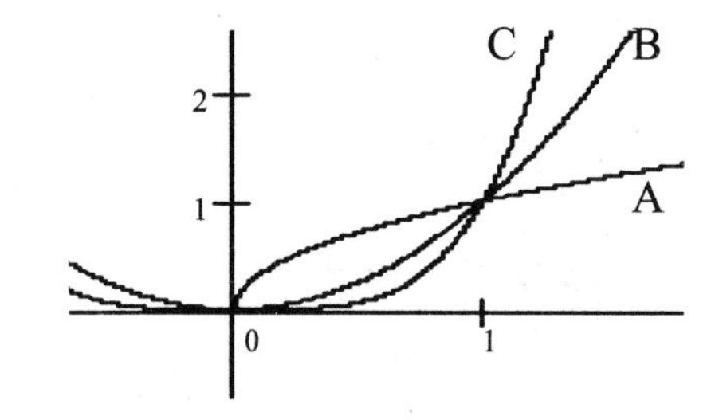

3. Match the following functions to the graphs. Explain briefly how you arrived at your answer.

   (i) $y = x^{1/2}$ (ii) $y = x^{1/3}$

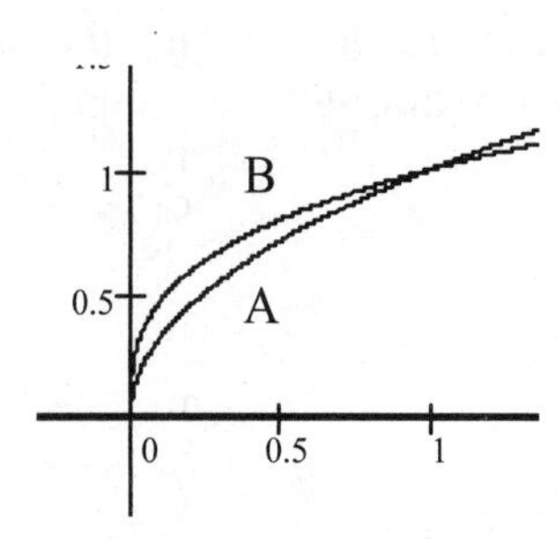

4. Find a formula for the power function f if $f(1) = 3$ and $f(4) = 6$.

Solutions to Practice Problems

1. (a) p is even because the graph is symmetric about the y-axis; (b) p is negative because the y-axis is a vertical asymptote; (c) $k = 3$ because $y = 3$ when $x = 1$.
2. A ↔ (iii) because the graph is concave down; B ↔ (i) and C ↔ (ii) because $x^4 > x^2$ for large values of x and $x^4 < x^2$ for $0 < x < 1$.
3. (i) ↔A and (ii) ↔ B, because $x^{1/2} < x^{1/3}$ for $0 < x < 1$.
4. $f(x) = kx^p$ since f is a power function; $f(1) = k(1^p) = 3$, so $k = 3$; $f(4) = 3 \cdot 4^p = 6$, so $4^p = 2$, and so $p = ½$. The formula is $f(x) = 3x^{1/2}$.

MASTERING CONCEPTS AND SKILLS

Use the √ , ?, * system on the exercises assigned by your instructor for this section.

Assigned Problems:

√

?

*

9.2 POLYNOMIAL FUNCTIONS

READING YOUR TEXTBOOK: Read Section 9.2, pp. 396-400.

As you read:

- Remember what you learned about power functions in Section 9.1. In particular, recall how the *sign* of the coefficient and whether the exponent is *odd* or *even* affect the long-run behavior of the graph.

- Learn the terminology associated with polynomials. In particular, know the meaning of all the words in **boldface** in the box on p. 398 and what is meant by the "zeros" of a polynomial. (See p. 399)

- Study the paragraph beginning at the bottom of p. 398 to understand why *the leading term determines the long-run behavior of a polynomial.* (See box, p. 399). Figures 9.15, 9.16, and 9.17, and Table 9.8 illustrate this behavior.

- Notice in Example 3, p. 399, how thinking about the long-run behavior helps you draw other conclusions about the polynomial.

REVIEWING THE BASICS

You should be able to:

- Determine whether or not a function is a polynomial.

- Determine the degree of a polynomial.

- Determine the long-run behavior of a polynomial by looking at its leading term.

Practice Problems

1. Determine which of the following functions are polynomials. State the degree and the long-run behavior of the polynomial functions.
   (a) $f(x) = 3x - 5x^3 + 7$

   (b) $f(x) = \dfrac{5}{x^2} + 2x - 8$

   (c) $f(x) = (x-1)(x+3)x^2$
2. For each polynomial function graphed below, state whether the degree is odd or even, and whether the leading coefficient is positive or negative.

(a)

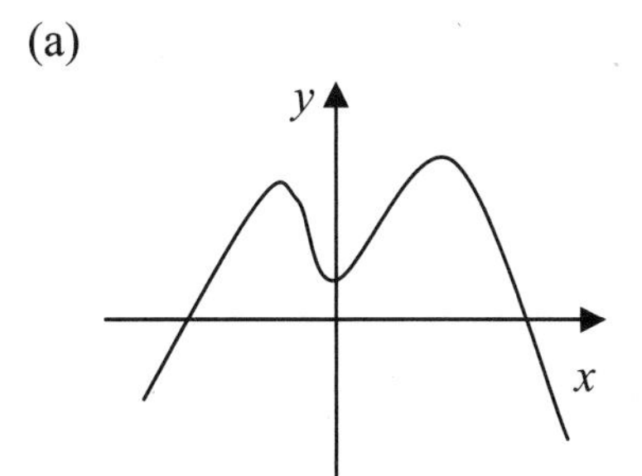

(b)

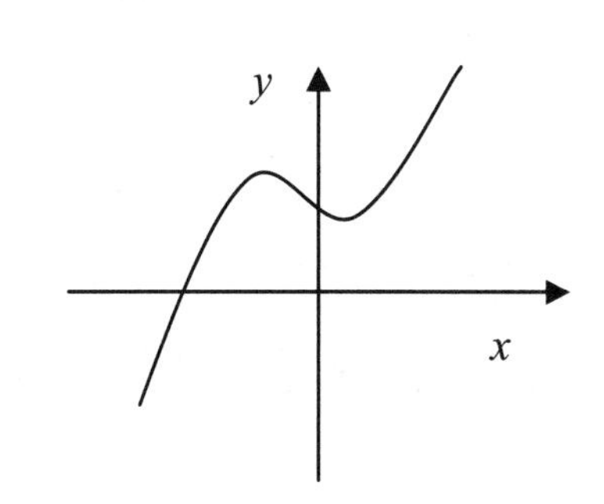

Solutions to Practice Problems

1. (a) polynomial; degree 3; on a large scale, $f(x)$ resembles $y = -5x^3$, which takes on large negative values as x takes on large positive values, and large positive values as x takes on large negative values. (b) not a polynomial; the term $5x^{-2}$ has a negative exponent. (c) polynomial; $f(x) = x^4 + 2x^3 - 3x^2$, so the degree is 4. On a large scale, $f(x)$ resembles the function $y = x^4$, which takes on large positive values as x grows large (either positive or negative).
2. (a) the degree is even, the leading coefficient is negative; (b) the degree is odd, the leading coefficient is positive.

MASTERING CONCEPTS AND SKILLS

Use the √ , ?, * system on the exercises assigned by your instructor for this section.

Assigned Problems:

√

?

---

*

9.3 THE SHORT-RUN BEHAVIOR OF POLYNOMIALS

READING YOUR TEXTBOOK: Read Section 9.3, pp. 402-406.

As you read:

- Notice that polynomials with the same leading coefficient will have the same long-run behavior, but they may differ in the *number of zeros* and the *number of turning points*. (See Example 1, p. 402.)

- Learn the important correspondence between *linear factors* and *zeros* of a polynomial. (See the first box, p. 404.)

- Learn the effect of *repeated linear factors* on the graph as it crosses or touches the x-axis, and what is meant by a *multiple zero*. . (See box, p. 405, and Figures 9.22 and 9.23.)

- Learn how the number of zeros and the number of turning points are related to the degree of the polynomial. (See box, middle of p.404.)

- Study carefully Example 4, p. 405. Notice how you use the x-intercepts (zeros) to write the factors of the polynomial, and why the factor $x - 3$ must appear to an even power. Notice also how you use the point (0,-3) to find the stretch factor k.

REVIEWING THE BASICS

You should be able to:

- Without using a calculator, draw a rough sketch of a polynomial given in factored form. Your graph should show the y-intercept, the x-intercepts, the effect of any multiple zeros, and the long-run behavior.

- Give a possible formula for a polynomial if you are given its graph.

Practice Problems

1. Without using a calculator, draw a rough sketch of the graph of $f(x) = -(x + 12)^2 (x - 20)$. You need not use the same scale on the x- and y-axes.

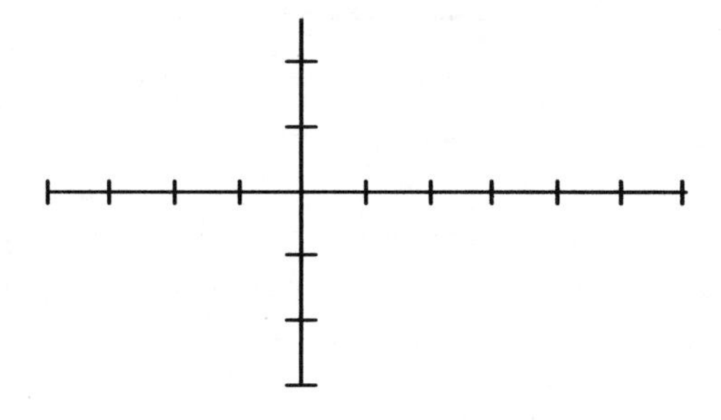

2. Give a possible formula for the polynomials having the following graphs.

(a)

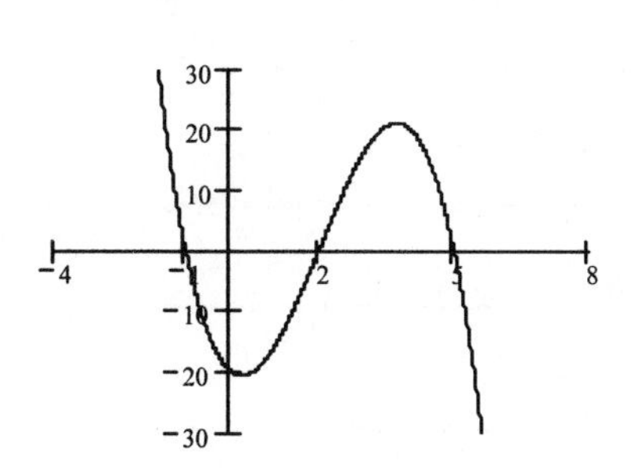

(b)

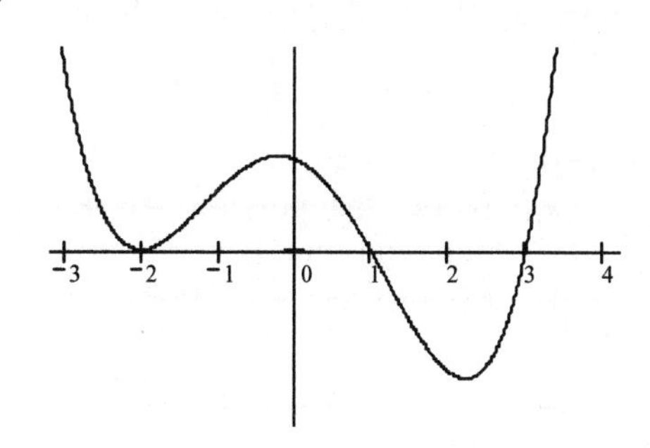

Solutions to Practice Problems

1. The x-intercepts are –12 and 20. Because the factor $x + 12$ appears to an even power, the graph touches the x-axis and turns but does not cross. The leading coefficient is –1, so the graph resembles $y = -x^3$ on a large scale.

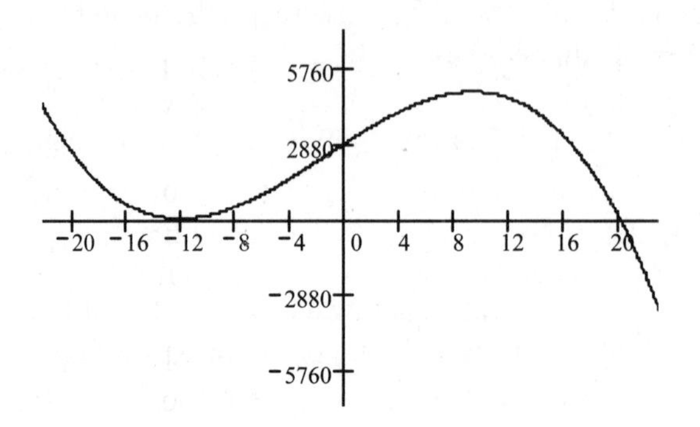

2. (a) $f(x) = k(x + 1)(x - 2)(x - 5)$. $f(0) = k(1)(-2)(-5) = -20$ if $k = -2$. So a possible formula is $f(x) = -2(x + 1)(x - 2)(x - 5)$. (b) A possible formula is $f(x) = (x + 2)^2(x - 1)(x - 3)$. Any positive stretch factor would give another possible formula, since no scale is given on the y-axis.

MASTERING CONCEPTS AND SKILLS

Use the √ , ?, * system on the exercises assigned by your instructor for this section.

Assigned Problems:

√

?

---

*

9.4 RATIONAL FUNCTIONS

READING YOUR TEXTBOOK: Read Section 9.4, pp. 409-413.

As you read:

- Learn what a *rational function* is. (See box, p. 410.) The opening discussion (pp. 409-410) describes a practical situation which gives rise to a rational function.
- Learn how to determine the long-run behavior of a rational function. (See box, p. 411, and Examples 1 and 2.) The evaluations in Example 1 will help you understand what it means to say that the line $y = 1$ is a horizontal asymptote.
- Remember from the previous section that polynomial graphs do not have horizontal asymptotes; recall what you have learned about the long-run behavior of polynomials. Notice also that rational functions do not have horizontal asymptotes when the degree of the numerator is larger than the degree of the denominator; see for example the function whose graph is given in Figure 9.28.

REVIEWING THE BASICS

You should be able to:

- Determine the long-run behavior of a rational function by looking at the quotient of its leading terms.
- Illustrate asymptotic behavior of a function with appropriate evaluations.
- Interpret asymptotic behavior of functions that model physical situations.

Practice Problems

1. Describe the long-run behavior of each of the following functions:

$$f(x) = \frac{x^3+2}{x^2+3} \qquad g(x) = \frac{x+2}{x^2+3} \qquad h(x) = \frac{x^3+2}{x^3+3}$$

2. Let $f(x) = \dfrac{3x+1}{x-2}$.

(a) Complete the following table for x-values close to 2. What happens to the values of $f(x)$ as x approaches 2 from the left? from the right?

x	1.5	1.9	1.99	2	2.01	2.1	2.5
$f(x)$							

(b) Complete the following tables. What happens to the values of $f(x)$ as x takes very large positive values? as x takes very large negative values?

x	5	10	100	1000
$f(x)$				

x	-5	-10	-100	-1000
$f(x)$				

(c) Give equations for the horizontal and vertical asymptotes of f.

(d) Find the x- and y-intercepts, and sketch the graph of f.

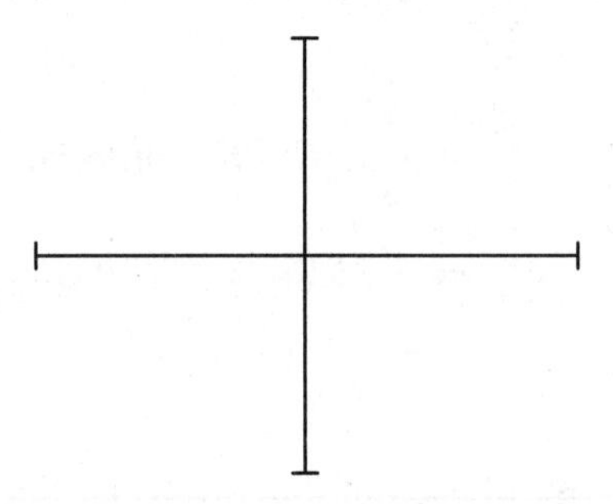

Solutions to Practice Problems

1. For large positive or negative values of x, $f(x) \approx \frac{x^3}{x^2} = x$, $g(x) \approx \frac{x}{x^2} = \frac{1}{x} \approx 0$, and

$h(x) \approx \frac{x^3}{x^3} = 1$.

2. (a)

x	1.5	1.9	1.99	2	2.01	2.1	2.5
$f(x)$	-11	-67	-697		703	73	17

The values of $f(x)$ are negative and are getting arbitrarily large in absolute value as x approaches 2 from the left. The values of $f(x)$ are getting arbitrarily large as x approaches 2 from the right.

(b)

x	5	10	100	1000
$f(x)$	5.3	3.9	3.07	3.007

x	-5	-10	-100	-1000
$f(x)$	2	2.4	2.93	2.993

The values of $f(x)$ are approaching 3.

(c) Horizontal: $y = 3$; vertical: $x = 2$.

(d) For the y-intercept, compute $f(0) = ½$; for the x-intercept set $3x + 1 = 0$; so $x = -⅓$.
To see the asymptotic behavior, we use one scale.

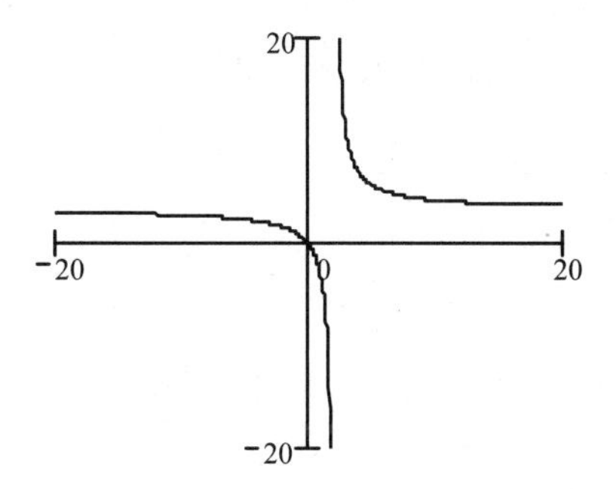

To see the behavior near the intercepts, we use another scale

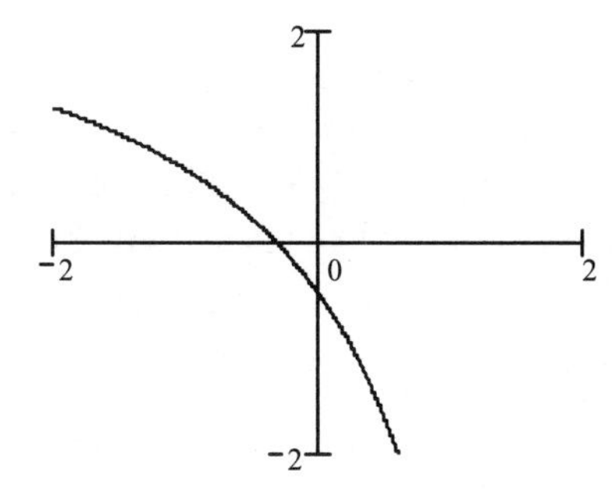

MASTERING CONCEPTS AND SKILLS

Use the √ , ?, * system on the exercises assigned by your instructor for this section.

Assigned Problems:

√

?

---

*

9.5 THE SHORT-RUN BEHAVIOR OF RATIONAL FUNCTIONS

READING YOUR TEXTBOOK: Read Section 9.5, pp. 415-419.

As you read:

- Remember the important fact about fractions: A fraction is equal to zero if its numerator equals zero (and its denominator does not). Use this fact when finding the *zeros of a rational function*. (See Example 1 p. 416.)

- Learn the facts summarized in the box, p. 417. Be sure to notice that the rational function is assumed to be reduced to lowest terms. (See p. 419 for an example of what can happen when both numerator and denominator have a common factor—that is, have the same zero.)

- Study Example 2 carefully. Notice how the important features of the graph—intercepts, horizontal asymptotes, vertical asymptotes, and the behavior of the graph near vertical asymptotes—let you draw a rough sketch without using a calculator. Knowing what the graph should look like can help you find an appropriate viewing window when using the calculator or computer.

- Notice in Example 3 how you use the x-intercepts to get the factors of the numerator and use the vertical asymptotes to get the factors of the denominator. Pay special attention to how the behavior of the graph tells you if the factor is repeated, and whether the exponent is odd or even.

- Note the important example at the top of p. 418 and Figure 9.37 which shows that a graph can intersect a horizontal asymptote. The graph of the function in Example 3 also meets its horizontal asymptote, the x-axis.

REVIEWING THE BASICS

You should be able to:

- Find the zeros and vertical asymptotes of a rational function that is written in factored form. Describe the long-run behavior and sketch the graph.
- Find a possible formula for a rational function by analyzing its graph.

Practice Problems

1. For each of the following rational functions, find all zeros and vertical asymptotes. Describe the behavior of the graph near each vertical asymptote, and describe the long-run behavior. Then sketch a graph without using a calculator.

   (a) $y = \dfrac{x+2}{x-5}$

   (b) $y = \dfrac{x+2}{(x-5)^2}$

2. The following graph is a translation of the graph of $y = 1/x$.
   (a) Find a possible formula that represents the graph.
   (b) Write the formula as the quotient of two linear polynomials.
   (c) Find the coordinates of the intercepts.

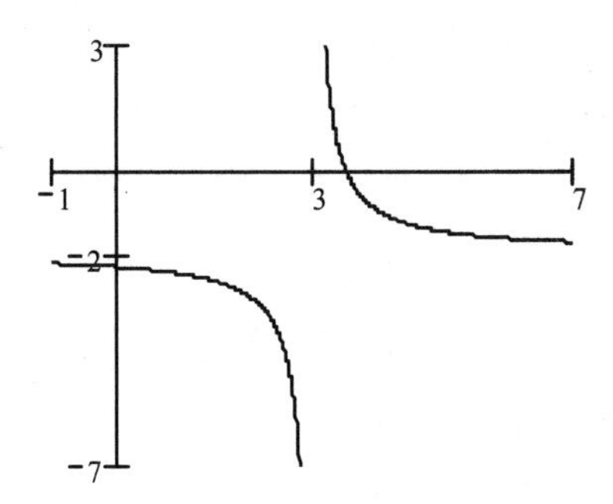

Solutions to Practice Problems

1. (a) $y = 0$ when $x + 2 = 0$, so the x-intercept is $x = -2$; $x - 5 = 0$ when $x = 5$, so the line $x = 5$ is a vertical asymptote; for large values of x, $y \approx \frac{x}{x} = 1$, so the line $y = 1$ is a horizontal asymptote.

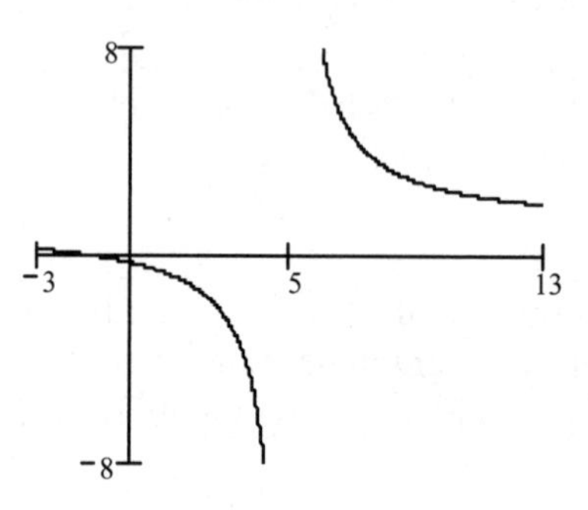

(b) $y = 0$ when $x = -2$; the line $x = 5$ is a vertical asymptote; near the line $x = 5$, the graph resembles $y = \frac{7}{(x-5)^2}$, so $y \to +\infty$ as $x \to 5$; for large values of x, $y \approx \frac{x}{x^2} = \frac{1}{x}$, so the x-axis is a horizontal asymptote.

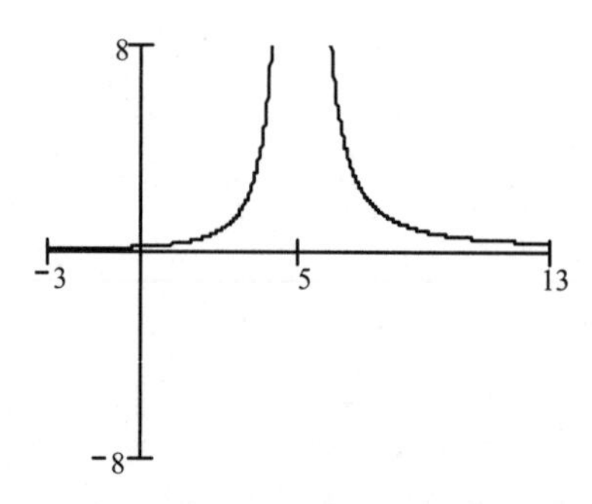

2. (a) The line $x = 3$ is a vertical asymptote, so we expect the denominator to have a factor of $x - 3$. There are no other vertical asymptotes, so we don't look for any other factors in the denominator. The horizontal asymptote is 2 units under the horizontal asymptote for $\frac{1}{x-3}$, so we shift the graph of $y = \frac{1}{x-3}$ down 2 units. A possible formula is $y = \frac{1}{x-3} - 2$; (b) $y = \frac{1-2(x-3)}{x-3} = \frac{-2x+7}{x-3}$; (c) the x-intercept is $x = 7/2$, and the y-intercept is $y = -7/3$.

MASTERING CONCEPTS AND SKILLS

Use the √ , ?, * system on the exercises assigned by your instructor for this section.

Assigned Problems:

√

?

*

9.6 COMPARING POWER, EXPONENTIAL, AND LOG FUNCTIONS

READING YOUR TEXTBOOK: Read Section 9.6, pp. 423-426.

As you read:

- Notice in Example 1, p. 423, that the long-term growth of a power function is affected more by the exponent than by the coefficient. The power function with the larger exponent eventually dominates.
- Study the comparison of the power function x^4 and the exponential function 2^x to see that the exponential function dominates. (See Figure 9.47.) Figure 9.48 shows that even the initially slower growing function 1.005^x eventually dominates the power function.
- Learn the summary that generalizes these examples (box, p. 424).
- Compare decreasing exponential functions to power functions with negative exponents (box, p. 425). The comparison is illustrated in Figures 9.49 and 9.50.
- Learn the comparison between power functions such as $x^{1/2}$ and $x^{1/3}$, and logarithm functions, which grow very slowly. The power functions are eventually larger. (See Figure 9.51.)

REVIEWING THE BASICS

You should be able to:

- Compare the long-term behavior of power, exponential, and log functions.

Practice Problems

1. How many solutions are there to the equation $\ln(x) = \dfrac{x}{10}$? Find them using your graphing calculator.
2. The graphs of $y = x^2$ and $y = 1.3^x$ meet in the first quadrant at the point where $x \approx 1.165$. Check this by graphing these functions in the window [0,10]x[0,10]. Explain how you know that the graphs must have another point of intersection in the first quadrant. Now graph both functions in the window [0,50]x[0,2500], and find the other point of intersection. Show the graphs in both windows on the axes provided. Are there any other solutions to $x^2 = 1.3^x$?

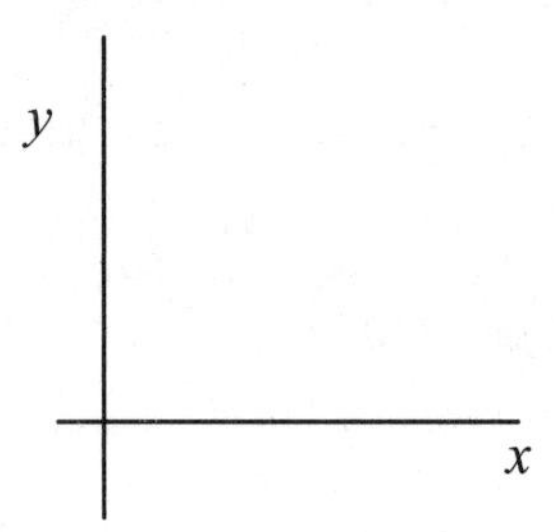

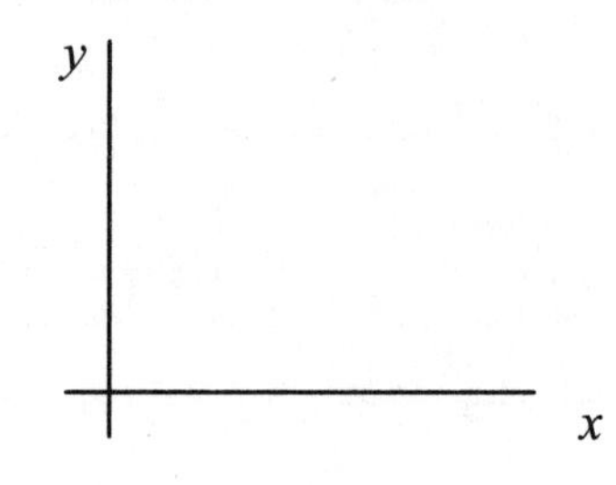

Solutions to Practice Problems

1. Graph both $y = \ln(x)$ and $y = \dfrac{x}{10}$ in the viewing window $0 \le x \le 5,\ 0 \le y \le 1$, and we can see that the graphs cross at about $x = 1.118$. The graph of $y = \ln(x)$ is above the graph of $y = \dfrac{x}{10}$ when $x = 5$, but because a linear function eventually dominates the natural log there must be another point of intersection. Experimenting with different windows lets us see the second point occurs when $x = 35.771$... All solutions must be positive because the domain of the ln function just includes positive numbers.

2. The graphs must meet again because the graph of the exponential function 1.3^x must eventually be above the graph of the power function x^2. The other solution is $x = 24.331$

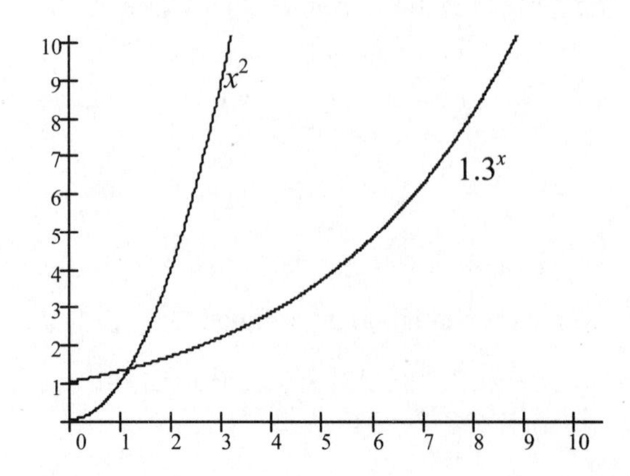

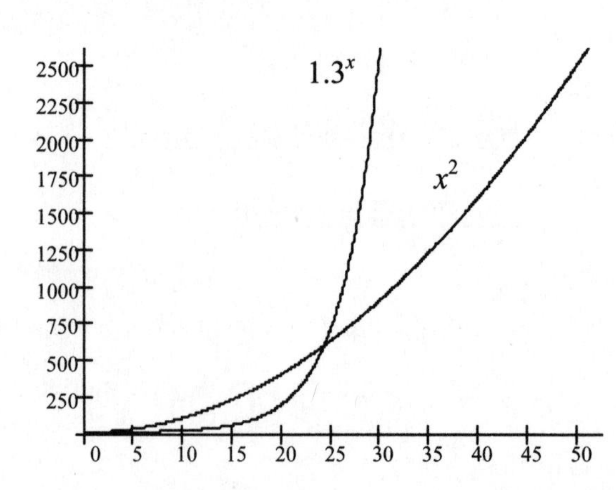

This time there is a third solution; the graph on the left above shows that near 0, $x^2 < 1.3^x$. But the graph of $y = x^2$ is symmetric with respect to the *y*-axis, so we know there is another point where the graphs meet; $x \approx -0.8898$.

MASTERING CONCEPTS AND SKILLS

Use the √ , ?, * system on the exercises assigned by your instructor for this section.

Assigned Problems:

√

?

---

*

9.7 FITTING EXPONENTIALS AND POLYNOMIALS TO DATA

READING YOUR TEXTBOOK: Read Section 9.7, pp. 428-430.

As you read:

- Be sure to do the calculations in the examples on your own calculator or computer. The numbers your calculator produces may be slightly different from the ones in the text.
- Review Section 1.6, **Fitting Linear Functions to Data**, and the homework exercises you did for that section to refresh your memory about linear regressions.
- Review the discussion on p. 47 of the text about correlation coefficients.
- Notice that the text uses three different types of functions (exponential, power, and linear) to model the same data; see Figure 9.59.

REVIEWING THE BASICS

You should be able to:

- Use the regression feature of your calculator to fit a specific type of curve to a given set of data.
- Compare two different models for the same data, and decide which one better represents the data.

Practice Problems

1. Use the data given below to answer the questions that follow.

x	2.07	3.25	4.68	5.91	7.34
y	18.75	27.93	43.11	58.73	98.26

(a) If you model this data with a straight line $y = mx + b$, what are m and b? What is the correlation coefficient?

(b) If you model this data by an exponential function, $y = Ae^{kx}$, what are A and k? What is the correlation coefficient?

(c) If you model this data by a power function, $y = Ax^p$, what are A and p? What is the correlation coefficient?

(d) Which of these seems to fit the data best?

Solutions to Practice Problems

1. (a) $m = 14.46$ and $b = -17.87$. The correlation coefficient is .965

(b) $A = 10.07$ and $k = 0.307$; $y = 10.07\, e^{0.307x}$. The correlation coefficient is 0.999.

(c) Here $A = 6.82$ and $p = 1.26$, so $y = 6.82\, x^{1.26}$ The correlation coefficient is 0.983.

(d) By comparing the correlation coefficients, it appears that the exponential function is the best fit.

MASTERING CONCEPTS AND SKILLS

Use the √ , ?, * system on the exercises assigned by your instructor for this section.

Assigned Problems:

√

?

---

*

CHAPTER TEN

VECTORS AND MATRICES

10.1 VECTORS

READING YOUR TEXTBOOK: Read Section 10.1, pp. 448-454.

As you read:

- Note that it takes two quantities to specify a vector: *magnitude* and *direction*.
- Be aware also of a new word that is introduced: *scalar*. This term is often used when vectors are being discussed, but it is not a new concept; a scalar is just a number.
- Be careful with notation; w stands for a number, whereas $\vec{w}$ stands for a vector. Also, $\|\vec{w}\|$ stands for the length, or magnitude, of $\vec{w}$.
- Study Figure 10.3, p. 449. It shows the first property in the second box on p. 454: the order in which you add two vectors doesn't matter.
- Learn how a vector is drawn and described. A vector is drawn as a line segment with an arrowhead on one end; the other end of the vector is called its tail. The geometric description of how you add vectors is given in Figure 10.5, p. 450.
- Remember the difference between the picture for adding vectors and the picture for subtracting vectors; compare Figures 10.5 (addition) and 10.8 (subtraction). Read the description of how to draw the arrow in each of these figures.
- Carefully follow the computations in Examples 2 and 3. Make sure you understand both the computation of magnitude and the computation of direction in each example.
- Envision the geometry of scalar multiplication; spend some time thinking about the statements in the first box on p. 454.
- Don't be overwhelmed by the five statements in the second box on p. 454. The last sentence after the box sums it up; the rules are what you expect them to be. Just be careful about how you combine objects; for example, you can't add a scalar and a vector.

REVIEWING THE BASICS

You should be able to:

- Recognize whether a quantity is a vector or a scalar.
- Given a picture of two vectors $\vec{u}$ and $\vec{v}$, draw the vectors $\vec{u}+\vec{v}$ and $\vec{u}-\vec{v}$.
- Given a vector $\vec{v}$ and a scalar k, draw the vector $k\,\vec{v}$.
- Given the magnitude and direction of two vectors, use the law of cosines to find the length of their sum and the law of sines to find the direction of their sum.

Practice Problems

1. Identify each of the following quantities as a vector or a scalar.
   (a) The force used to push a crate.
   (b) The age of a student.
   (c) The speed of a car.
   (d) The velocity of a car.

2. Given the two vectors $\vec{u}$ and $\vec{v}$ in the following sketch, draw and label as $\vec{q}$ the vector $\vec{u}+\vec{v}$. Draw and label as $\vec{w}$ the vector $\vec{u}-\vec{v}$. Write a sentence describing the relationship between $\vec{w}$ and the vector $\vec{v}-\vec{u}$.

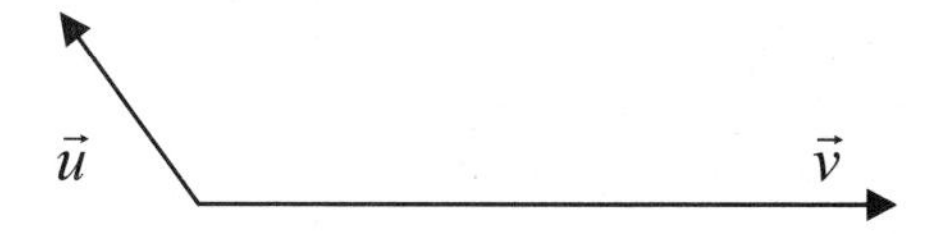

3. Given the vector $\vec{u}$ in the sketch, draw and label the following vectors: $\vec{R}=2\vec{u}$; $\vec{S}=\frac{1}{3}\vec{u}$; and $\vec{V}=-\vec{u}$.

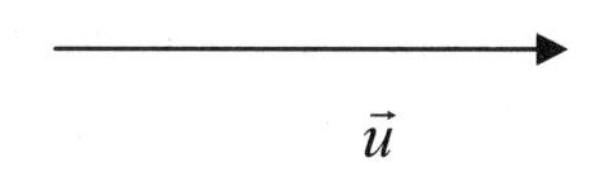

4. For the vectors in the following sketch, $\|\vec{u}\|=8$ and $\|\vec{v}\|=5$. Find the magnitude and direction of the vector $\vec{u}+\vec{v}$.

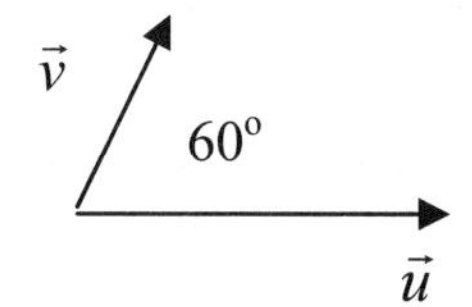

Solutions to Practice Problems

1. (a) vector; (b) scalar; (c) scalar; (d) vector.
2. The vector $\vec{v} - \vec{u}$ goes from the tip of $\vec{u}$ to the tip of $\vec{v}$, so it is just the opposite of $\vec{u} - \vec{v}$, which goes from the tip of $\vec{v}$ to the tip of $\vec{u}$.

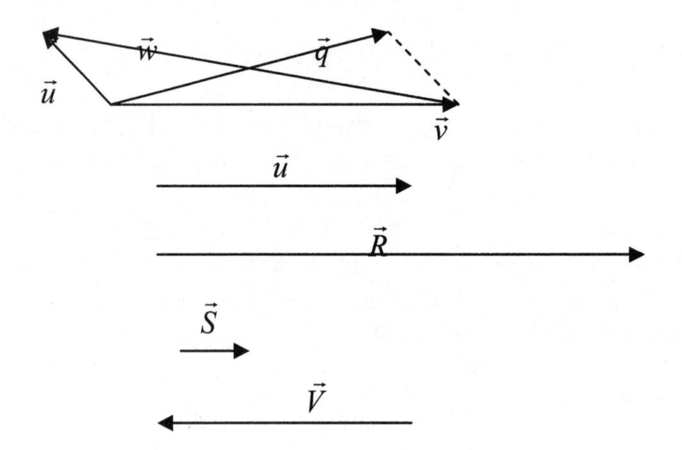

3.

4. Let x be the length of $\vec{u}+\vec{v}$. Using the law of cosines,
   $x^2 = 8^2 + 5^2 - 2 \cdot 8 \cdot 5 \cdot \cos 120^\circ$
   $= 64 + 25 - 80(-1/2)$
   $= 64 + 25 + 40 = 129.$

   Thus the magnitude of $\vec{u}+\vec{v}$ is $\sqrt{129} = 11.4$.

   Now we can use the Law of Sines to find the angle θ between $\vec{u}+\vec{v}$ and $\vec{u}$.

   $$\frac{\sin\theta}{5} = \frac{\sin 120^0}{x}; \text{ so } \sin\theta = 5 \cdot \frac{0.886}{11.4} = 0.380. \text{ Thus } \theta = \sin^{-1}(0.380) = 22.3^\circ.$$

   The vector $\vec{u}+\vec{v}$ has length 11.4 and forms an angle of 22.3° counterclockwise from $\vec{u}$.

MASTERING CONCEPTS AND SKILLS

Use the √ , ?, * system on the exercises assigned by your instructor for this section.

Assigned Problems:

√

?

---

*

10.2 THE COMPONENTS OF A VECTOR

READING YOUR TEXTBOOK: Read Section 10.2, pp. 456-459

As you read:

- Watch for the introduction of the two special vectors $\vec{i}$ and $\vec{j}$. Note that $\vec{i}$ is the vector from the origin to the point (1,0) and $\vec{j}$ is the vector from the origin to the point (0,1).
- Become familiar with the phrase "Resolve a vector into its components."
- Understand the formula in the box on p. 457.
- Understand the formula in the box on p. 459.

REVIEWING THE BASICS

You should be able to:

- Find the components of a vector given its magnitude and the angle the vector makes with the positive x-axis.
- Given two points P and Q in the plane, find the components of the vector from P to Q using the $\vec{i}$, $\vec{j}$ notation.
- Find a unit vector in any given direction.

Practice Problems

1. Find the components of the vector with length 12 that makes an angle of 135° counterclockwise with the positive x-axis.

2. Find the components of the vector from the point $P = (8, -2)$ to the point $Q = (5,9)$.

3. (a) Find the components of the unit vector that makes an angle of 210° counterclockwise with the positive x-axis.

   (b) Find the unit vector that points in the same direction as the vector in Practice Problem 2 of this section.

Solutions to Practice Problems

1. The vector is $(12\cos 135^0)\,\vec{i} + (12\sin 135^0)\,\vec{j} = -6\sqrt{2}\,\vec{i} + 6\sqrt{2}\,\vec{j} = -8.49\vec{i} + 8.49\,\vec{j}$.
2. The vector is $(5-8)\vec{i} + (9-(-2))\vec{j} = -3\vec{i} + 11\vec{j}$.
3. (a) $(\cos 210^0)\,\vec{i} + (\sin 210^0)\,\vec{j} = -0.866\vec{i} - 0.5\,\vec{j}$. (b) The length of the vector $-3\vec{i} + 11\vec{j}$ is $\sqrt{(-3)^2 + 11^2} = \sqrt{130}$. The unit vector in the same direction as a given vector is found by dividing the given vector by its length. Thus the unit vector we want is $(-3\vec{i} + 11\vec{j})/\sqrt{130} = \dfrac{-3}{\sqrt{130}}\vec{i} + \dfrac{11}{\sqrt{130}}\vec{j}$.

MASTERING CONCEPTS AND SKILLS

Use the √ , ?, * system on the exercises assigned by your instructor for this section.

Assigned Problems:

√

?

---

*

10.3 APPLICATIONS OF VECTORS

READING YOUR TEXTBOOK: Read Section 10.3, pp. 461-466.

As you read:

- Note the use of the term *position vector*; this vector always starts at the origin.See the last sentence in this section, right before the Exercises on p. 466.

- When presented with, for example, the notation (5,7), be sure the read the context to decide whether the notation is being used to represent a point or a vector.

REVIEWING THE BASICS

You should be able to:

- Represent vectors with more than two components.

- Add vectors with the same number of components.
- Multiply a vector by a scalar.

Practice Problems

1. If $\vec{u} = (3,2,-4,8)$ and $\vec{v} = (7,-2,5,1)$, compute:
   (a) $\vec{u} + \vec{v}$

   (b) $2\ \vec{u}$

2. Find a vector seven units long that when added to the vector (20,5) results in a vector whose second coordinate is 0.

Solutions to Practice Problems

1. (a) $\vec{u} + \vec{v} = (3{+}7, 2-2, -4{+}5, 8{+}1) = (10,0,1,9)$.
   (b) $2\ \vec{u} = (2{\cdot}3, 2{\cdot}2, 2{\cdot} -4, 2{\cdot}8) = (6, 4, -8, 16)$.
2. Let $\vec{u}$ be the vector seven units long. Then $\vec{u} = (7 \cos \theta, 7 \sin \theta)$, and we want $5 + 7 \sin \theta$ to be zero. Thus $5 + 7 \sin \theta = 0$, or $\sin \theta = -5/7$. Now there are two solutions to this equation for $0 \le \theta < 2\pi$, or equivalently $0° \le \theta < 360°$. Using degrees, we have $\theta = 225.6°$ or $314.4°$. Thus there are two such vectors: $\vec{u} = (7 \cos 225.6°, 7 \sin 225.6°) = (-4.90, -5)$ and $\vec{u} = (7 \cos 314.4°, 7 \sin 314.4°) = (4.90, -5)$.

.MASTERING CONCEPTS AND SKILLS

Use the √ , ?, * system on the exercises assigned by your instructor for this section.

Assigned Problems:

√

?

---

*

10.4 THE DOT PRODUCT

READING YOUR TEXTBOOK: Read Section 10.4, pp. 468-472.

As you read:

- Understand how to compute the dot product of two vectors (see box p. 468); remember that your answer will always be a number.
- Note the properties in the box on p. 469. The middle two are just what you expect to be true. The first property is the Law of Cosines in a new form. Also, given that the angle between a vector and itself is 0, the first property implies the last property.
- Note that if the dot product of two vectors is 0, then the two vectors are perpendicular. If their dot product is positive, the angle between them is acute. If their dot product is negative, the angle between them is obtuse.

REVIEWING THE BASICS

You should be able to:

- Compute the dot product of two vectors.
- Use the dot product to discuss the angle between two vectors.

Practice Problems

1. (a) Compute $(7,5) \cdot (2,4)$.

   (b) Compute $(2\vec{i} + 3\vec{j}) \cdot (5\vec{i} - 6\vec{j})$.

2. Find a number c so that the vector $(2, -1, c)$ is perpendicular to the vector $(3, 7, 4)$.

3. Find the angle between the vectors $(-5, 9, 1)$ and $(2, 3, 8)$. (Remember that the angle between two vectors is always between 0° and 180°.)

Solutions to Practice Problems

1. (a) (7, 5) ·(2, 4) = 7·2 + 5·4 = 14 + 20 = 34. (b) $(2\vec{i}+3\vec{j})\cdot(5\vec{i}-6\vec{j})$ = 2·5 +3· − 6 = − 8.
2. We want (2, − 1, c) ·(3, 7, 4) = 0. So 6 – 7 + 4c = 0. So 4c = 1, and c = ¼.
3. (− 5, 9, 1) ·(2, 3, 8) = − 10 + 27 + 8 = 25. $\|(-5,9,1)\| = \sqrt{25+81+1} = \sqrt{107}.$ $\|(2,3,8)\| = \sqrt{4+9+64} = \sqrt{77}.$ Thus 25 = $\sqrt{107}\sqrt{77}$ cos θ, where θ is the angle between the vectors. So cos θ = 0.275, and $\theta = \cos^{-1}(0.275) = 74.0°$.

MASTERING CONCEPTS AND SKILLS

Use the √ , ?, * system on the exercises assigned by your instructor for this section.

Assigned Problems:

√

?

---

*

10.5 MATRICES

READING YOUR TEXTBOOK: Read Section 10.5, pp. 474-480.

As you read:

- Learn the definition of *matrix* as a rectangular grid of numbers; see p. 474.
- Learn how to multiply a matrix by a scalar, and how to add and subtract matrices; see Example 1 and the first box on p.476.
- Be familiar with the properties of scalar multiplication and matrix addition; see second box p. 476.
- Carefully follow the population discussion on pp.476-477 to see how matrices can be used to write 2 linear equations as one matrix equation. See box on p. 478 and Examples 3 and 5.
- Learn how to multiply an $n \times n$ matrix, A, with an n-dimensional vector, $\vec{u}$, forming $A\vec{u}$. See the bottom of p. 478.
- Learn the correspondence between $A\vec{u}$ and a system of n linear equations. See Examples 2-5.

REVIEWING THE BASICS

You should be able to:

- Add two matrices, and tell when two matrices cannot be added.
- Multiply a matrix by a scalar.
- Express a system of n equations in n variables in the form $A\vec{u}$.
- Perform the matrix multiplication $A\vec{u}$.

Practice Problems

1. Perform the indicated operation, or state why it cannot be done.

$$A = \begin{pmatrix} 2 & 5 \\ -4 & 7 \end{pmatrix}, \quad B = \begin{pmatrix} -1 & 3 & 9 \\ 0 & -2 & 5 \end{pmatrix}, \quad C = \begin{pmatrix} 5 & -1 & -6 \\ 2 & 7 & 4 \end{pmatrix}$$

(a) $3A$

(b) $B + C$

(c) $A + B$

2. Compute $\begin{pmatrix} 2 & 0 & -1 \\ 5 & 3 & -2 \\ 8 & 4 & 7 \end{pmatrix}\begin{pmatrix} 9 \\ 6 \\ -4 \end{pmatrix}$

1.

2. Write the system $\begin{matrix} 3x + 2y = u_1 \\ 5x - 6y = u_2 \end{matrix}$ as $A\vec{v} = \vec{u}$

Solutions to Practice Problems

1. (a) $\begin{pmatrix} 6 & 15 \\ -12 & 21 \end{pmatrix}$ (b) $\begin{pmatrix} 4 & 2 & 3 \\ 2 & 5 & 9 \end{pmatrix}$ (c) Cannot be done; B has an extra column of numbers.

2. $$\begin{pmatrix} 18+0+4 \\ 45+18+8 \\ 72+24-28 \end{pmatrix} = \begin{pmatrix} 22 \\ 71 \\ 68 \end{pmatrix}$$

3. $$\begin{pmatrix} 3 & 2 \\ 5 & -6 \end{pmatrix}\begin{pmatrix} x \\ y \end{pmatrix} = \begin{pmatrix} u_1 \\ u_2 \end{pmatrix}$$

MASTERING CONCEPTS AND SKILLS

Use the √ , ?, * system on the exercises assigned by your instructor for this section.

Assigned Problems:

√

?

---

*

CHAPTER ELEVEN

SEQUENCES AND SERIES

11.1 SEQUENCES

READING YOUR TEXTBOOK: Read Section 11.1, pp. 488-491.

As you read:

- Learn the general terminology: *sequence* and *terms* of the sequence. See top of p.488
- Understand what makes a sequence an *arithmetic* sequence. The definition is at the top of p.489. The box on p.489 gives a formula for the nth term.
- Understand what makes a sequence a *geometric* sequence. The definition is just before Example 6, p. 490. The box on p. 491 gives a formula for the nth term.

REVIEWING THE BASICS

You should be able to:

- Recognize an arithmetic sequence, and identify the common difference.
- Calculate an arbitrary term in an arithmetic sequence
- Recognize a geometric sequence, and identify the common ratio..
- Calculate an arbitrary term in a geometric sequence.

Practice Problems

1. Identify which of the following are arithmetic sequences, and determine d for each arithmetic sequence.
   (a) 1,2,4,8,16,32,.....

   (b) 12,10,8,6,......

   (c) 80,83,86,88,.......

   (d) 12,16,20,24,28,......

2. Given the arithmetic sequence 3, 7,...., what is the next term? What is the 9th term? What is the nth term?

3. Identify which of the following are geometric sequences, and determine the common ratio for each geometric sequence.
   (a) 2, 6, 18, 54,......
   (b) 4, -4, 4, -4,.......
   (c) 3, 5, 7, 9,.....
   (d) 40, 20, 10, 5,.......

4. Given the geometric sequence 4, 12,.... what is the next term? What is the 9th term? What is the nth term?

Solutions to Practice Problems

1. (a) not arithmetic; (b) arithmetic, d = -2; (c) not arithmetic; (d) arithmetic, d = 4.

2. The next term is 11; the 9th term is 35; the nth term is $3+4(n-1)$.

3. (a) geometric, r = 3; (b) geometric, r = -1; (c) not geometric; (d) geometric, r = ½.

4. The next term is 36; the 9th term is $4 \cdot 3^8 = 26244$; the nth term is $4 \cdot 3^{(n-1)}$.

MASTERING CONCEPTS AND SKILLS

Use the √, ?, * system on the exercises assigned by your instructor for this section.

Assigned Problems:

√

?

---

*

11.2 DEFINING FUNCTIONS USING SUMS: ARITHMETIC SERIES

READING YOUR TEXTBOOK: Read Section 11.2, pp. 493-498.

As you read:

- Learn the general terminology; *series* and n^{th} partial sum. See p. 493.
- Be sure to distinguish between an arithmetic sequence and an arithmetic series.
- Learn the formula for the nth partial sum of an arithmetic series in the box on p.495. The formula is obtained on p.495. The formula for S_n is just *n* times the average of the first term and the nth term.
- Study the meaning of the summation notation between Examples 3 and 4 on p.496. Notice how it is used in Examples 4 and 5.
- Examples 6 and 7 show how to use the formula for S_n. Follow the computations carefully.

REVIEWING THE BASICS

You should be able to:

- Recognize an arithmetic series, and be able to calculate S_n.
- Use the sigma notation to write S_n.

Practice Problems

1. Decide which of the following series are arithmetic series. For each arithmetic series, find the common difference *d*, and compute the indicated sum using the formula for S_n.

   (a) $5 + 10 + 15 + 20 + 25 + 30$

   (b) $64 + 32 + 16 + 8 + 4$

   (c) $2 + 5 + 8 + 11 + ... + 122$

2. Write the following arithmetic series using sigma notation; compute the sum.

   (a) $2 + 6 + 10 + 14 + 18 + 22$

(b) 25 + 22 + 19 + 16 + 13 + 10 + 7

(c) 4 + 6 + 8 +. . . + 126

3. Calculate the following sums.

(a) $\sum_{i=0}^{15} 6$

(b) $\sum_{n=1}^{10} (3n)$

(c) $\sum_{n=0}^{12} (5+2n)$

Solutions to Practice Problems

1. (a) arithmetic, $d = 5$, $S_6 = \frac{1}{2}6(5+30) = 105$. (b) not arithmetic; (c) not arithmetic;

   (d) arithmetic, $d = 3$; set $122 = 2 + (n - 1)3$, and get $n = 41$, and

   $S_{41} = \frac{1}{2}41(2+122) = 2542$.

2. (a) $\sum_{i=1}^{6} (2+4(i-1))$; note with $i = 1$, the first term is 2; the common difference is 4, so that is the coefficient of $(i - 1)$, and we stop with $i = 6$ because that gives the value of the last term; $2 + 4 \cdot 5 = 22$.

   (b) The series starts with 25 and d = -3, so the sum will look like $\sum_{i=1}^{?} (25-3(i-1))$. The last term is 7, and $25 - 3(i - 1) = 7$ means $i = 7$. So the answer is

   $\sum_{i=1}^{7} (25-3(i-1))$.

   (c) The series starts with 4 and $d = 2$, so $126 = 4 + 2(n - 1)$, and $n = 62$.

   $S_{62} = \frac{1}{2}62(4+126) = 4030$

3. (a) This sum is $6 + 6 + 6 + \ldots + 6$ where there are 16 terms. Note that the sum starts with $i = 0$, and there are 16 numbers in the list 0,1,2,...,15. So the answer here is 96. To use the formula on p.484, note that we use $d = 0$, $n = 16$, and $a_1 = 6$. So $S_{16} = (1/2)(16)(6 + 6) = 96$. (b) Here the initial term $a_1 = 3$, $d = 3$, and $n = 10$. So the sum is $(1/2)(10)(3 + 30) = 165$. (c) The initial term is 5, $d = 2$, and the number of terms is 13. The sum is $(1/2)(13)(5 + 29) = 221$.

MASTERING CONCEPTS AND SKILLS

Use the √ , ?, * system on the exercises assigned by your instructor for this section.

Assigned Problems:

√

?

---

*

11.3 FINITE GEOMETRIC SERIES

READING YOUR TEXTBOOK: Read Section 11.3, pp. 500-504.

As you read:

- Follow the arithmetic on p.500 and in Example 1 carefully. Verify the results with your own calculator.
- Follow the steps on p.502 that lead to the formula for S_n which is given in the box.
- Understand that the relation of a geometric series to a geometric sequence is the same as the relation of an arithmetic series to an arithmetic sequence.
- Be sure you understand Example 3 before reading Example 4, which is a continuation of Example 3.

REVIEWING THE BASICS

You should be able to:

- Recognize a finite geometric series and be able to compute S_n.
- Use the sigma notation to write a geometric series.

Practice Problems

1. Decide which of the following series are geometric series, and find S_n for each one that is.
   (a) 1 + 2 + 3 + 4 + ... + 100

   (b) 5 + 10 + 20 + 40 + 80 + 160 + 320

   (c) 64 + 32 + 16 + 8 + 4

   (d) $4-2+1-\frac{1}{2}+\frac{1}{4}-\frac{1}{8}$

2. Write the following geometric series using sigma notation.

   (a) 1 + 2 + 4 + 8 + 16 +. . . + 512

   (b) $8+2+\frac{1}{2}+\frac{1}{8}+\frac{1}{32}+\frac{1}{128}$

3. Find the sum of each of the following finite geometric series.

   (a) $6+4+\frac{8}{3}+\frac{16}{9}+\frac{32}{7}$

   (c) $10-5+\frac{5}{2}-\frac{5}{4}+...+10(-1/2)^8$

Solutions to Practice Problems

1. (a) not geometric; (b) geometric, $r = 2$; $S_7 = \frac{5(1-2^7)}{1-2} = 635$ (c) geometric, $r = ½$, $S_5 = \frac{64(1-0.5^5)}{1-0.5} = 124$ (d) geometric, $r = -½$, $S_6 = \frac{4(1-(-0.5)^6)}{1-(-0.5)} = 2.625$

2. (a) The initial term is 1; the ratio $r = 2$, and the last term is of the form $1 \cdot 2^i$ where $i = 9$. Thus an answer is $\sum_{i=0}^{9} 1(2)^i$, or more simply $\sum_{i=0}^{9} 2^i$. (b) The initial term is 8, the ratio $r = 1/4$, and the last term is of the form $8(1/4)^n$ where $n = 5$, so the answer is $\sum_{n=0}^{5} 8(1/4)^n$.

3. (a) This is a finite geometric series with initial term 6 and ratio 2/3. The last term is of the form $6(2/3)^4$, so the sum is given by $\dfrac{6(1-(2/3)^5)}{1-2/3} = 15.630$.
(b) This is a finite geometric series whose initial term is 10 and the ratio is –1/2. The last term is $10(-1/2)^8$, so the sum is $\dfrac{10(1-(-1/2)^9)}{1-(-1/2)} = 6.680$.

MASTERING CONCEPTS AND SKILLS

Use the √, ?, * system on the exercises assigned by your instructor for this section.

Assigned Problems:

√

?

*

11.4 INFINITE GEOMETRIC SERIES

READING YOUR TEXTBOOK: Read Section 11.4, pp. 505-509.

As you read:

- Notice that the opening discussion in Section 11.4 is a continuation of the discussion in Section 11.3
- Memorize the formula for the sum of an infinite geometric series in the box on p. 506 carefully; it is only valid for $|r| < 1$. The formula for the finite sum from Section 11.3 is valid for all $r \neq 1$.
- Learn the definition of *present value* (see first box p.508) and how to compute it (second box p.508).

REVIEWING THE BASICS

You should be able to:

- Recognize an infinite geometric series, and find its sum.
- Calculate the present value of a future payment.

Practice Problems

1. Write each of the following geometric series using sigma notation, and compute its sum.

(a) $6+2+\frac{2}{3}+\frac{2}{9}+\frac{2}{27}+...$

(b) $6+4+\frac{8}{3}+\frac{16}{9}+...$

2. Find the sum of each of the following geometric series.

(a) $\sum_{i=0}^{\infty} 10(-1/2)^i$

(b) $\sum_{n=2}^{\infty} 3(1/5)^n$

Solutions to Practice Problems

1. (a) This is an infinite series whose initial term is 6 and the ratio is 1/3, so the answer is

$$\sum_{n=0}^{\infty} 6(1/3)^n = \frac{6}{1-\frac{1}{3}} = 9$$

(b) This series also has 6 as its initial term, but now $r = 2/3$, so the answer is given by $\sum_{i=0}^{\infty} 6(\frac{2}{3})^i = \frac{6}{1-2/3} = 18$.

2. (a) The initial term is 10, and $r = -1/2$, so the sum is $\dfrac{10}{1-\dfrac{-1}{2}} = \dfrac{20}{3}$

   (b) Be careful here; the initial term is $3(1/5)^2 = 3/25$; the ratio is $r = 1/5$. The answer is $\dfrac{3/25}{1-1/5} = \dfrac{3}{20}$.

MASTERING CONCEPTS AND SKILLS

Use the √ , ?, * system on the exercises assigned by your instructor for this section.

Assigned Problems:

√

?

*

CHAPTER TWELVE

PARAMETRIC EQUATIONS AND CONIC SECTIONS

12.1 PARAMETRIC EQUATIONS

READING YOUR TEXTBOOK: Read Section 12.1, pp. 516-523.

As you read:

- Notice the introduction of a new word-- *parameter.* The word *parameter* is just another word for *variable* and is commonly used when position on a curve is given as a function of time, as described in the first paragraph on p.517.

- Remember that you have seen parameters before; θ is a parameter in the definition of $\cos\theta$ and $\sin\theta$ on p.251.

- Understand that "to eliminate the parameter" means to write the equation of the curve in Cartesian coordinates.

- Learn from Example 3 that replacing t by $\frac{1}{2}t$ decreases the speed of a particle along a path. Similarly, replacing t by $2t$ would increase the speed of the particle. Example 4 shows that replacing t by $-t$ would cause the particle to travel in the opposite direction.

- Check with your instructor about using your graphing calculator or other technology to draw graphs given by parametric equations.

- Carefully work through Example 5 before attempting Exercises 13-16.

REVIEWING THE BASICS

You should be able to:

- Draw a graph of a parametrically defined curve.

- Given a parametrically defined curve, eliminate the parameter and write a Cartesian equation for the curve; that is, write an equation for the curve using only x and y.

- Write several different parameterizations for the graph $y = f(x)$.

Practice Problems

1. Sketch the curve given by each of the following parametric equations. Indicate the direction of the curve. Eliminate the parameter t and write the curve in the Cartesian equation $y = f(x)$ if y is a function of x.

(a) $x = 3t - 2 \quad y = \cos 2t$ Draw the part of the graph given by $0 \le t \le 3$.

(b) $x = 2 + 3 \cos t \quad y = 1 - 4 \sin t$ Draw the portion of the graph given by $0 \le t \le 2\pi$.

2. Give two different parameterizations of the curve $y = 2^x$ that go from the point (0,1) and end at the point (3,8).

Solutions to Practice Problems

1. (a) The graph is as follows; the curve goes from left to right.

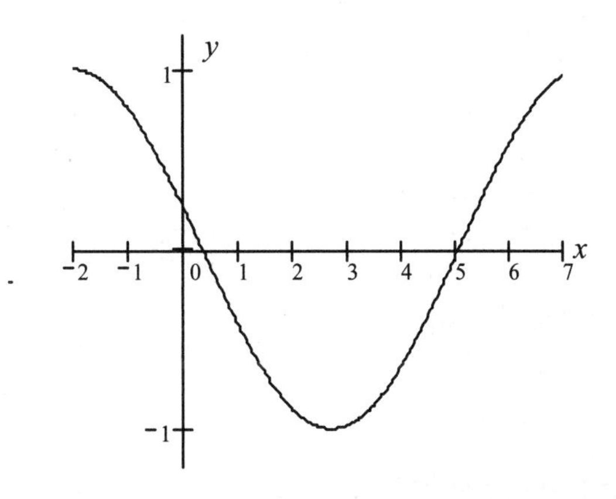

To eliminate the parameter t, we can solve the first equation for t in terms of x: $t = \dfrac{x+2}{3}$. Substituting into the second equation gives $y = \cos 2\left(\dfrac{x+2}{3}\right)$.

(b) In these equations, we don't want to end up with a complicated expression involving sin composed with $\cos^{-1}$. Instead we recall that $\cos^2 t + \sin^2 t = 1$. We solve the first equation for cos t and the second equation for sin t. We get $\cos t = \dfrac{x-2}{3}$ and $\sin t = \dfrac{y-1}{-4}$. Our equation is $\left(\dfrac{x-2}{3}\right)^2 + \left(\dfrac{y-1}{-4}\right)^2 = 1$, or, since we are squaring, $\left(\dfrac{x-2}{3}\right)^2 + \left(\dfrac{y-1}{4}\right)^2 = 1$, the equation of an ellipse. Thus the curve fails the vertical line test and is not the graph of a function. We can write the functions for the top and bottom half of the ellipse. Start with $\dfrac{(x-2)^2}{9} + \dfrac{(y-1)^2}{16} = 1$ to get $16(x-2)^2 + 9(y-1)^2 = 144$; so $9(y-1)^2 = 144 - 16(x-2)^2$, or

$y-1=\frac{\pm\sqrt{144-16(x-2)^2}}{3}$. The two functions are $y=1+\frac{\sqrt{144-16(x-2)^2}}{3}$ and $y=1-\frac{\sqrt{144-16(x-2)^2}}{3}$. The graph is shown in the following sketch.

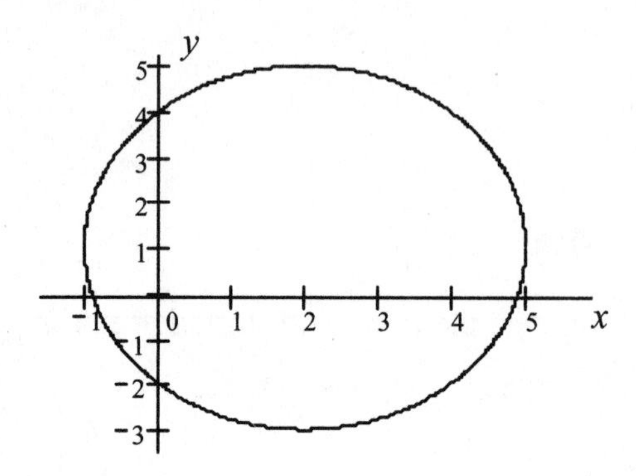

2. We just go the opposite way from the way we solved Practice Problem 1(a). The simplest equation for x in terms of t would simply be $x = t$, so $y = 2^t$. One parameterization then is given by $x = t$, $y = 2^t$, $0 \leq t \leq 3$. Now just make up x as some other function of t so that x goes from 0 to 3; for example, $x = 3t$, $0 \leq t \leq 1$. Then $y = 2^x = 2^{3t}$, and a second parameterization is $x = 3t$, $y = 2^{3t}$, $0 \leq t \leq 1$.

MASTERING CONCEPTS AND SKILLS

Use the √ , ?, * system on the exercises assigned by your instructor for this section.

Assigned Problems:

√

?

---

*

12.2 IMPLICITLY DEFINED CURVES AND CIRCLES

READING YOUR TEXTBOOK: Read Section 12.2, pp. 525-528.

As you read:

- Review the method of "completing the square" given in "*Tools for Chapter 5*", p. 239.

- Check with your instructor about using your graphing calculator or other technology to draw graphs given by parametric equations.

- Become familiar with the term *conic section.* A conic section is either a parabola, a circle, an ellipse (See Section 12.3) or a hyperbola (See Section 12.4).

- Notice the role that $\cos^2 t + \sin^2 t = 1$ plays in the parameterization of circles.

- Remember that a circle implicitly defines two functions because it fails the vertical line test; $y^2 = 16$ has two solutions.

REVIEWING THE BASICS

You should be able to:

- Read off the center and radius of a circle when its equation is in standard form.

- Given an equation for a circle, rewrite the equation in standard form by completing the square.

- Parameterize a circle given its center and radius.

Practice Problems

1. Identify each of the following equations as a circle or not. Give a verbal description of each circle, and then sketch the graph.
   (a) $(x-4)^2 + (y-3)^2 = 49$

   (b) $(x-3)^2 - (y-2)^2 = 16$

(c) $x^2 + y^3 + 2y = 10$

(d) $5x^2 - 3x - 8y = 12$

(e) $x^2 + y^2 + 4x - 4y + 25 = 16$

2. (a) Write the implicit equation for the circle centered at $(5, -2)$ with radius 8.

(b) Write parametric equations for the circle described in part (a).

Solutions to Practice Problems

1. (a) Circle, centered at (4, 3), radius 7.

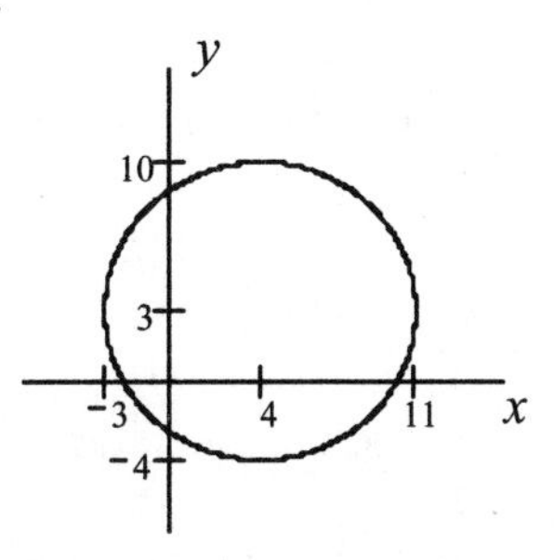

(b) Not a circle, the equation is the difference of squares, not the sum.

(c) Not a circle; there is a y^3 in the equation. Circles are only given by certain quadratic equations.

(d) Not a circle; you should recognize that this is the equation of a parabola.

(e) Not a circle; the coefficients of x^2 and y^2 are equal, but when we complete the square to rewrite in standard form, we get $(x+2)^2+(y-2)^2=-1$.

2. (a) $(x-5)^2+(y+2)^2=64$

(b) $x=5+8\cos(t)$, $y=-2+8\sin(t)$

MASTERING CONCEPTS AND SKILLS

Use the √ , ?, * system on the exercises assigned by your instructor for this section.

Assigned Problems:

√

?

---

*

12.3 ELLIPSES

READING YOUR TEXTBOOK: Read Section 12.3, pp. 529-530.

As you read:

- Learn how to represent an ellipse parametrically; see box p. 529.
- Learn how to represent an ellipse implicitly; see box p. 530.
- Learn the definitions of major axis and minor axis; these are given just before Example 2 on p. 530.
- Notice that a circle is a special case of an ellipse.
- Remember that an ellipse is one of the conic sections.
- Remember that an ellipse implicitly defines two functions because it fails the vertical line test; $y^2 = 16$ has two solutions.

REVIEWING THE BASICS

You should be able to:

- Read off the center and lengths of the major and minor axes of an ellipse when its equation is in standard form.
- Given an equation for an ellipse, rewrite the equation in standard form by completing the square.
- Parameterize an ellipse given its center and lengths of it horizontal and vertical axes.

Practice Problems

1. Identify each of the following equations as an ellipse or not. Give a verbal description of each ellipse, specifying its center and lengths of its horizontal and vertical axes..

   (a) $(x-4)^2 + 4(y-3)^2 = 64$

   (b) $4(x-3)^2 - (y-2)^2 = 16$

(c) $3x^2 + y^3 + 2y = 10$

(d) $5x^2 - 3x - 8y = 12$

(e) $x^2 + 2y^2 - 6x + 16y + 16 = 25$

2. (a) Write the implicit equation for the ellipse centered at $(5, -2)$, whose horizontal axis has length 8 and whose vertical axis has length 20.

(b) Write parametric equations for the ellipse described in part (a).

3. Find the center and the lengths of the major and minor axes for the ellipse $\dfrac{(x+2)^2}{9}+\dfrac{(y-3)^2}{16}=1$. Sketch the graph.

Solutions to Practice Problems

1. (a) Ellipse, centered at (4, 3), horizontal axis has length $2\sqrt{64}=16$, vertical axis has length $2\sqrt{\dfrac{64}{4}}=8$.

   (b) Not an ellipse, the coefficients of x^2 and y^2 have different signs.

   (c) Not an ellipse; there is a y^3 in the equation. Ellipses are only given by certain quadratic equations.

   (d) Not an ellipse; you should recognize that this is the equation of a parabola.

   (e) Ellipse; the equation is $\dfrac{(x-3)^2}{25}+\dfrac{2(y+4)^2}{25}=1$. The length of the horizontal axis is 10, and the length of the vertical axis is $5\sqrt{2}$

2. (a) $\dfrac{(x-5)^2}{16}+\dfrac{(y+2)^2}{100}=1$

   (b) $x=5+4\cos(t), \quad y=-2+10\sin(t)$

3. The ellipse is centered at (-2, 3). The length of the horizontal axis is $2\sqrt{9}=6$, and the length of the vertical axis is $2\sqrt{16}=8$. Since the longer one is the major axis, the major axis has length 8 and the minor axis has length 6. The graph follows.

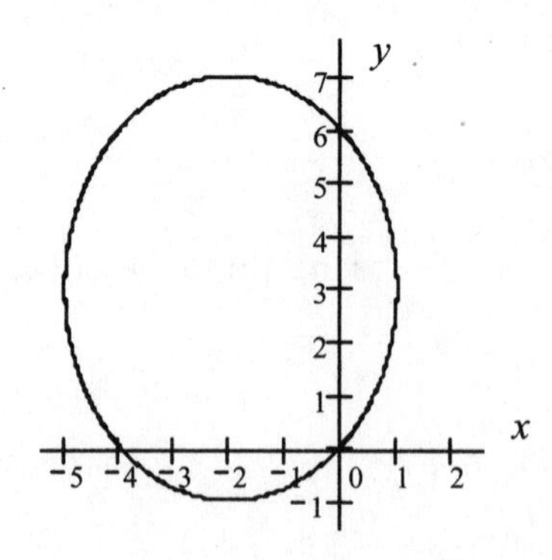

MASTERING CONCEPTS AND SKILLS

Use the √ , ?, * system on the exercises assigned by your instructor for this section.

Assigned Problems:

√

?

---

*

12.4 HYPERBOLAS

READING YOUR TEXTBOOK: Read Section 12.4, pp. 533-535.

As you read:

- Learn how to represent a hyperbola parametrically; see box p. 534.
- Notice how the trig identity $\sec^2 t - \tan^2 t = 1$ plays the same role in the parameterization of a hyperbola that $\cos^2 t + \sin^2 t = 1$ plays in the parameterization of circles and ellipses.
- Learn how to represent a hyperbola implicitly; see box p. 535.
- Recognize when a hyperbola opens left and right, or up and down.
- Learn how to find the asymptotes of a hyperbola.
- Remember that a hyperbola is one of the conic sections.
- Remember that a hyperbola implicitly defines two functions because it fails the vertical line test; $y^2 = 16$ has two solutions.

REVIEWING THE BASICS

You should be able to:

- Find the center and the vertices of a hyperbola. (See Example 1 p.534)
- Given an equation for a hyperbola, rewrite the equation in standard form by completing the square.
- Parameterize a hyperbola given its center and vertices.

Practice Problems

1. Find the center, vertices, and asymptotes for the hyperbola $\frac{(y-4)^2}{25}-\frac{(x+1)^2}{9}=1$.

   Sketch the graph.

2. Identify each of the following equations as a hyperbola or not. Sketch the graph of each hyperbola.

   (a) $(x-4)^2+(y-3)^2=49$

   (b) $\frac{(x-3)^2}{16}-\frac{(y-2)^2}{36}=1$

   (c) $x^2+y^3+2y=10$

(d) $5x^2 - 3x - 8y = 12$

(e) $2x^2 + y^2 + 4x - 4y + 6 = 4$

3. (a) Write the implicit equation for the hyperbola with vertices at (14, 4) and (-16, 4) so that the asymptotes have slopes 2/5 and –2/5..

(b) Write the parametric equations for the hyperbola described in part (a).

(c) Sketch the graph of this hyperbola.

Solutions to Practice Problems

1. The center of the hyperbola is (-1, 4), the vertices are $(-1,9)$ and $(-1,-1)$; note the hyperbola opens up and down. The asymptotes are $y-4=\frac{5}{3}(x+1)$ and $y-4=\frac{-5}{3}(x+1)$. The graph is below.

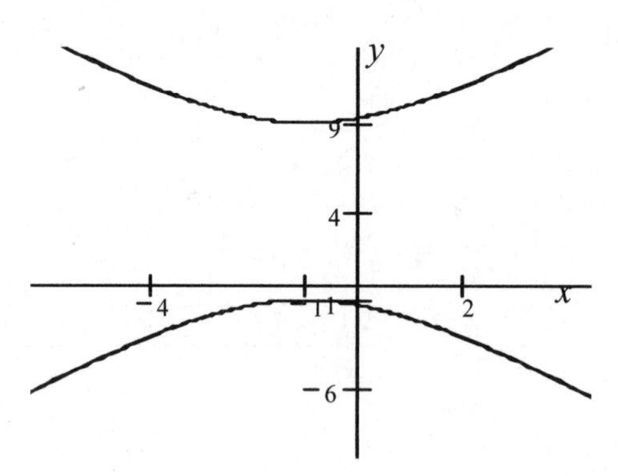

2. (a) Not a hyperbola; this is the equation of a circle.

   (b) Hyperbola, centered at (3, 2), opens left and right (since the coefficient of x^2 is positive and the coefficient of y^2 is negative).

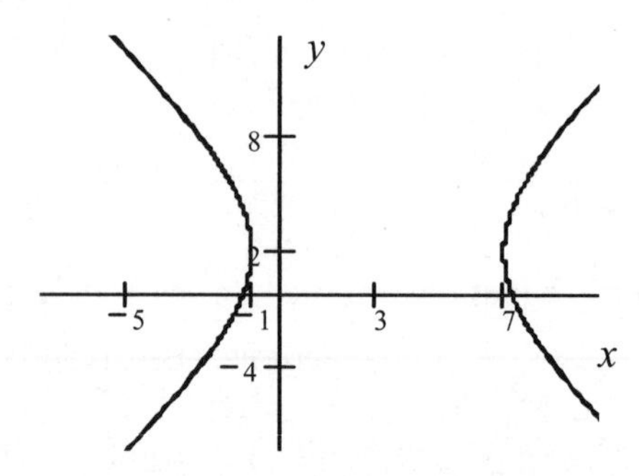

   (c) Not a hyperbola; hyperbolas involve quadratic expressions, not cubic.

   (d) Not a hyperbola; this is the equation of a parabola.

   (e) Not a hyperbola; this is the equation of an ellipse.

3. (a)Vertices with the same *y*-coordinate means that the hyperbola opens up left and right, so the coefficient of y^2 is negative. The center is halfway between the vertices, so the center is (-1, 4). The vertices are 30 units apart, so $2a = 30$, and $a = 15$. The slope of one the asymptotes is 2/5, so $\frac{b}{a}=\frac{b}{15}=\frac{2}{5}$; $b=6$. Answer: $\frac{(x+1)^2}{225}-\frac{(y-4)^2}{36}=1$.

   (b) $x = -1 + 15 \sec t$, $y = 4 + 6 \tan t$.

(c)

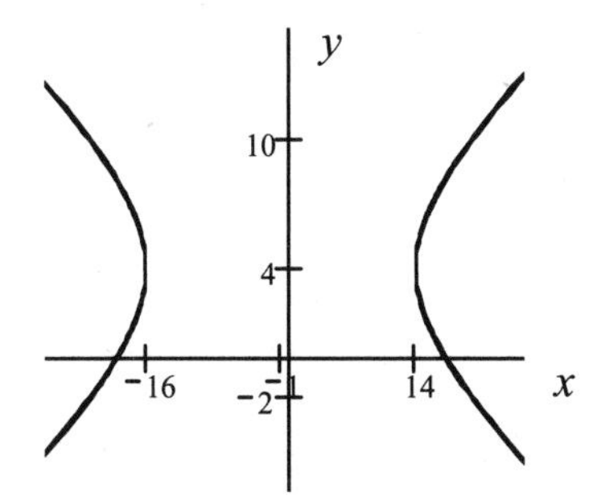

12.5 GEOMETRIC PROPRTIES OF CONIC SECTIONS

READING YOUR TEXTBOOK: Read Section 12.5, pp. 537-548

As you read:

- Learn the geometric definition of a circle. (First box, p.539)
- Learn the geometric definition of an ellipse, focal point, major axis, and minor axis. (Second box, p. 539)
- Note that in the solution to Example 1, part (e), the parameter t does not stand for time.
- Study Example 2 to learn how to find the foci using the implicit equation for the ellipse. (Box p.542)
- Learn the geometric definition of a hyperbola. (Box p.543)
- Study Example 4 to learn how to find the focal points of a hyperbola. (Box p.544)
- Learn the geometric definition of a parabola. (Box p.546)
- Study Example 7 to learn how to find the focus and directrix of a parabola. (Box p.547)
- Learn the reflective properties of the ellipse (Box p.542), hyperbola (Box p.545), and parabola (Box p.548)

REVIEWING THE BASICS

You should be able to:

- Identify which conic is associated with a given quadratic equation.
- Find center and focal points of an ellipse.
- Find vertices and focal points of a hyperbola.
- Find vertex, focus, and directrix of a parabola.
- Find an equation for an ellipse, hyperbola, or parabola from given geometric information.

Practice Problems

1. Find the vertex, focal point and directrix of the parabola $y = x^2 - 6x + 11$

2. Find an equation of the ellipse whose focal points are (-3, 0) and (3, 0) if the length of the minor axis is 8.

3. A beam of light starts at the point (15, 84) and travels down and to the left along a line whose slope is 3. It meets the hyperbola $\frac{x^2}{144} - \frac{y^2}{25} = 1$. Can you tell where the light beam meets the x-axis?

Solutions to Practice Problems

1. We complete the square to rewrite the equation as $y = (x-3)^2 + 11$, so the vertex is the point (3, 2). Using the notation in the box on p.547, we have $a = 1$, so the vertical distance from to the directrix and focus is $\frac{1}{4a} = \frac{1}{4}$. The directrix is the line $y = 1.75$, and the focus is the point (3, 2.25).

2. We have the center of the ellipse is the point (0, 0) and the major axis lies along the x-axis, so the equation of the ellipse is $\frac{x^2}{a^2} + \frac{y^2}{b^2} = 1$, with $a > b$. Also $2b = 8$, so $b = 4$. Next $c = \sqrt{a^2 - b^2}$, so $3 = \sqrt{a^2 - 16}$; $a^2 = 25$. The equation of the ellipse is $\frac{x^2}{25} + \frac{y^2}{16} = 1$

3. The light is traveling on the line $y - 84 = 3(x - 15)$, or $y = 3x + 39$. The x- intercept of this line is (-13, 0), which is where the light beam would meet the axis if the hyperbola were not there. Now the x-coordinates of the focal points of this hyperbola are given by $c = \pm\sqrt{144 + 25} = \pm 13$, so the focal points are (-13, 0) and (13, 0). Because the beam of light would have gone through (-13, 0), it is reflected through the other focus by the hyperbola. The light beam meets the x-axis at the point (13, 0).

MASTERING CONCEPTS AND SKILLS

Use the √ , ?, * system on the exercises assigned by your instructor for this section.

Assigned Problems:

√

?

*

12.6 HYPERBOLIC FUNCTIONS

READING YOUR TEXTBOOK: Read Section 12.6, pp. 552-555.

As you read:

- Learn the definitions of cosh x and sinh x (box, p. 552) and how these functions are pronounced (p.552).
- You may at first think it strange that the hyperbolic cosine is the name given to a function defined in terms of e^x and e^{-x}. But recall that Euler's formula (see p.344) made a connection between the exponential function and the circular trig functions. The material on p. 554-555 shows that the connection between cosh t, sinh t, and the hyperbola is similar to the connection between cos t, sin t, and the circle.
- Note that the definition of tanh x is analogous to the definition of tan x. (See box, p. 554.)

REVIEWING THE BASICS

You should be able to:

- Show that cosh x is even and sinh x is odd.
- State and use the identity relating cosh x and sinh x in the first box on p. 454.
- Parameterize one branch of a hyperbola using hyperbolic functions.

Practice Problems

1. What is the domain of $\tanh x$? Is $\tanh x$ an odd function, an even function, or neither? Justify your answers.

2. Describe the behavior of $\tanh x$ as $x \to \infty$ and as $x \to -\infty$

3. What should be the definition of $\operatorname{sech} x$? Use this definition to get an identity involving $\tanh x$ and $\operatorname{sech} x$.

4. Parameterize the branch of the hyperbola $16(y-2)^2 - 25(x-3)^2 = 400$ with $y > 2$.

Solutions to Practice Problems

1. Since $\cosh x > 0$ for all x, the domain of $\tanh x$ is all real numbers. Also, $\tanh x$ is an odd function, because $\tanh(-x) = \dfrac{\sinh(-x)}{\cosh(-x)} = \dfrac{-\sinh x}{\cosh x} = -\tanh x$.

2. As $x \to \infty$, $\tanh x = \dfrac{e^x - e^{-x}}{e^x + e^{-x}} \to \dfrac{e^x}{e^x} = 1$. (Also note that $\cosh x > \sinh x$ for all x implies $\tanh x < 1$ for all x.) As $x \to -\infty$, $\tanh x \to \dfrac{-e^{-x}}{e^{-x}} = -1$. (Note also that $|\cosh x| > |\sinh x|$ implies $-1 < \tanh x < 1$ for all x, so the range of $y = \tanh x$ is $-1 < y < 1$.)

3. We can define $\operatorname{sech} x$ to be $1/\cosh x$, analogous to $\sec x = 1/\cos x$. Motivated by $\sec^2 x - \tan^2 x = 1$, we try for an identity involving $\operatorname{sech}^2 x$ and $\tanh^2 x$; $\operatorname{sech}^2 x + \tanh^2 x = \dfrac{1}{\cosh^2 x} + \dfrac{\sinh^2 x}{\cosh^2 x} = \dfrac{1}{\cosh^2 x}(1+\sinh^2 x) = \dfrac{1}{\cosh^2 x}(\cosh^2 x) = 1$.

   So our answer is $\operatorname{sech}^2 x + \tanh^2 x = 1$.

4. We start by rewriting the equation in standard form: $\dfrac{(y-2)^2}{25} - \dfrac{(x-3)^2}{16} = 1$. To transform this to $\cosh^2(t) - \sinh^2(t) = 1$, first we want $\dfrac{(y-2)^2}{25} = \cosh^2(t)$, so either $y = 2 + 5\cosh(t)$, or $y = 2 - 5\cosh(t)$. Since the branch we want has $y > 2$, we choose $y = 2 + 5\cosh(t)$. Next we want $\dfrac{(x-3)^2}{16} = \sinh^2(t)$. So that the parameterization goes from left to right, we choose $x = 3 + 4\sinh(t)$. Use your graphing calculator to see what would happen if you choose $x = 3 - 4\sinh(t)$.

MASTERING CONCEPTS AND SKILLS

Use the √ , ?, * system on the exercises assigned by your instructor for this section.

Assigned Problems:

√

?

---

*

NOTES

NOTES

NOTES

NOTES